Corporate Contributors

Santee Cooper

South Carolina Electric and Gas Company

Duke Power Company

Elf Atochem North America, Inc.

Griffin Corporation

Monsanto Company

Union Camp Corporation

J.M. Malone and Son Enterprises

Keo Fish Farms, Inc.

The generous support provided by these contributors helped defray the printing cost of this publication. This savings has allowed the Department of Natural Resources to provide this publication to the public at a reduced cost. The authors extend their sincere appreciation to all the corporate contributors for their continued support of this Second Edition.

Aquatic and Wetland Plants of South Carolina

Cynthia A. Aulbach-Smith
Botanical Services of South Carolina

Steven J. de Kozlowski
South Carolina Department of Natural Resources

Algae section contributed by
Lawrence A. Dyck, *Clemson University*

South Carolina Department of Natural Resources
in cooperation with the
South Carolina Aquatic Plant Management Council

June 1996

Printed in the United States of America.

Illustrations by Victoria C. Hollowell

Original book design by R. Brian Murrell, RBM Design, Columbia, SC

Cover design by Diane S. Kennedy, SC Department of Natural Resources,
Columbia, SC

Aquatic and Wetland Plants of South Carolina
Second Edition
ISBN 0-9632821-1-5

Copies may be obtained from:

Publications Coordinator
Water Resources Division
South Carolina Department of Natural Resources
1201 Main Street, Suite 1100
Columbia, South Carolina 29201
(803) 737-0800

Table of Contents

Physiographic Provinces of South Carolina

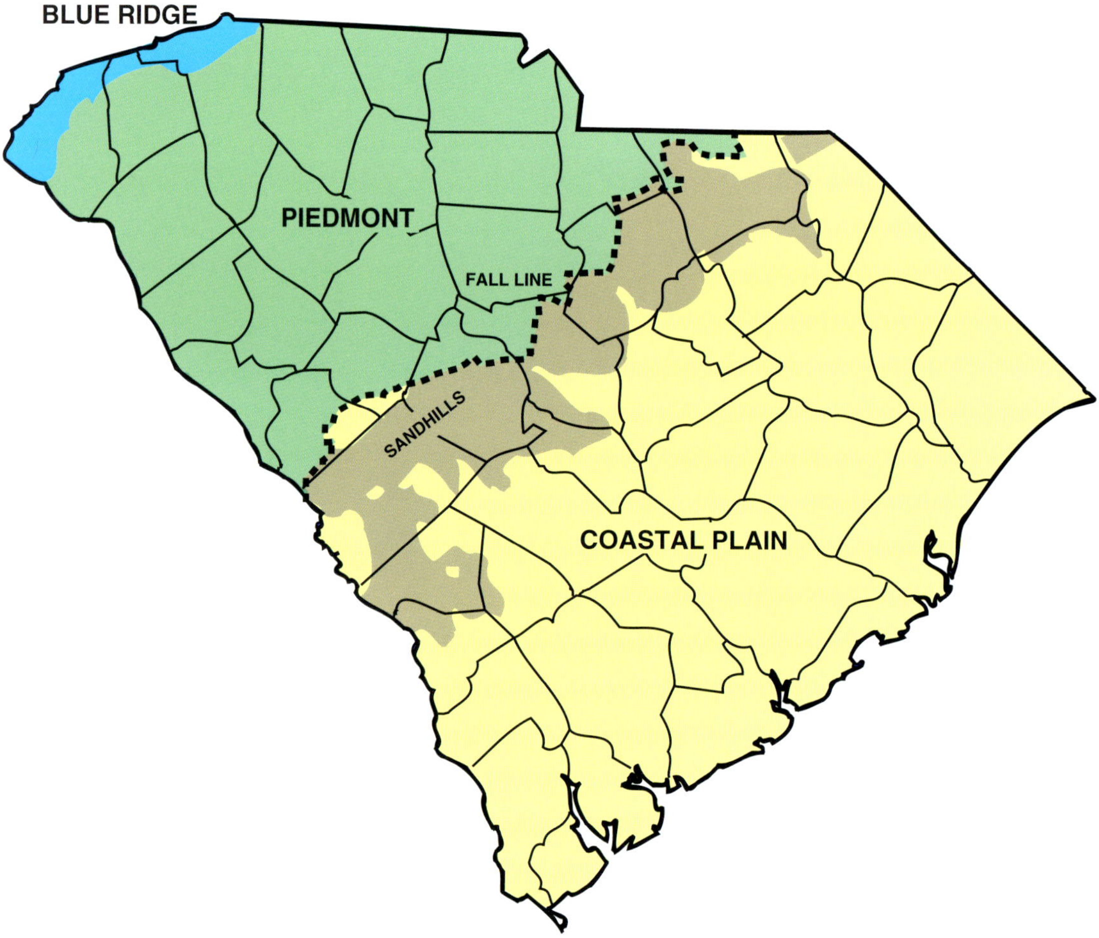

Introduction

South Carolina's surface water resources are vital to the state's environmental health and economic growth. Over 460,000 acres of public lakes and 11,000 miles of rivers provide an abundant supply of fresh water for municipal, industrial, and agricultural withdrawal purposes and electric power generation. They also support a variety of recreational activities including swimming, boating, and fishing. As residential development around our major lakes increases, interest in aquatic environments by the general public has grown.

Plants are an important component of these aquatic systems providing food, shelter, and reproductive habitat for fish and wildlife, and enhancing aesthetic appeal. Most plants found in the state's waters are beneficial native species with sufficient natural controls to limit their growth. However, a few exotic species, especially some recent introductions, lack natural controls, and may develop large populations that outcompete desirable native plants and adversely impact water use activities. These species require ongoing management efforts to help minimize their environmental and economic impacts.

In 1980, the South Carolina Aquatic Plant Management Council was established to provide statewide coordination of aquatic plant management activities on public waters. The Council recognized that public education and awareness of aquatic plants and their accurate identification were important for an effective statewide management program, especially in preventing the introduction and spread of noxious non-native species. Therefore in 1985 the Council approved the development of an aquatic plant identification manual for use by the general public and aquatic plant management personnel. Fortunately, the basis for a statewide manual had already been developed under a project supported by the Santee-Cooper River Basin Study Group entitled *Aquatic and Wetland Plants of the Santee Cooper Lake System, South Carolina* (Aulbach-Smith, 1986, unpublished). This manual includes much of the photographic material, illustrations, and written descriptions from that project, with the addition of information from subsequent statewide surveys and existing herbarium specimens.

Purpose and Scope

The purpose of this book is to provide an easy to use reference to accurately identify aquatic and selected wetland plants in South Carolina. While comprehensive scientific references are available for identifying aquatic plant species in the Southeast, they require a functional command of botanical terms and the ability to use dichotomous keys. This publication combines accurate botanical descriptions with detailed illustrations and colorful photographs in a "user friendly" design that allows for quick and easy identification. The result is a manual for the scientific community that complements existing publications and a useful and usable reference for the general public to easily identify aquatic and wetland plants throughout the state.

Most of the state's freshwater aquatic plants are included with the notable exceptions of trees, shrubs, and those wetland plants that are not obligate aquatic species. Close to 2,000 species are considered wetland plants in South Carolina. Therefore, adequate coverage of all wetland plants is beyond the scope of this work. Furthermore, since the focus is on freshwater aquatics, those plants restricted to brackish and salt water habitats are not included. Also omitted are some species that are actively managed in tidal freshwater impoundments for wildlife purposes but are rarely found outside of these managed areas.

Organization

The vascular plants are organized into four chapters based primarily on growth habit. A select group of algae that are frequently encountered in aquatic systems is presented in the last chapter. All plant descriptions are based on specimens collected in South Carolina. Within each chapter, plants are presented in phylogenetic order by family, and then alphabetically by genus and species within the family.

For the reader's convenience, each chapter is color coded so it can be readily identified. A plant description can therefore be found by simply selecting the chapter that includes plants possessing growth habits similar to the plant in question and then identifying the plant by using the pictures. Plants can also be located by referencing the index which includes both the common and scientific names. Page numbers referring to the primary treatment of each plant are indicated in the index with bold type. Although this book was written for use by nonbotanists, the written descriptions of each plant include botanical terms. Therefore, an illustrated glossary is included in the back of the book for reference.

State & Federal Noxious Weed Lists

Preventing the occurrence and spread of aquatic weed infestations in public waters can saves millions of public and private dollars each year in avoided economic losses and control costs. Because prevention is so important, state and federal laws have been adopted which prohibit the possession, importation, and distribution of specific exotic aquatic plant species known to cause problems.

The South Carolina Noxious Weed Act of 1971 provides the S.C. Department of Agriculture with far reaching powers to "seize, quarantine, treat, destroy,... or otherwise dispose of... any noxious weed" found within the state. To further deter persons from spreading nuisance aquatic weeds the law includes fines up to $500 and/or imprisonment up to one year.

The Federal Noxious Weed Act was passed in 1974 to provide authority to prevent the introduction of foreign weeds into the United States and to allow eradication of listed species before they become widespread. The U.S. Department of Agriculture, Animal and Plant Health Inspection Service (APHIS) is responsible for implementing and enforcing the Federal Noxious Weed Act.

The S.C. Department of Natural Resources and the Aquatic Plant Management Council coordinate with the S.C. Department of Agriculture and the U.S. Department of Agriculture to implement these laws as they relate to nuisance aquatic plants. Educating the public and the plant nursery industry about these laws and the potential problems associated with certain aquatic plants is an important component of the state's prevention efforts.

Aquatic plant species currently listed in the Federal and State Noxious Weed Acts as illegal to possess, import, and distribute in South Carolina are listed below. The S.C. Department of Natural Resources, Aquatic Plant Management Program staff should be notified regarding the occurrence of these species in public waters or their possession or sale within South Carolina.

Federal Noxious Weed List

Mosquito fern	*Azolla pinnata*
Rooted water hyacinth	*Eichhornia azurea*
Hydrilla	*Hydrilla verticillata*
Miramar weed	*Hygrophila polysperma*
Water spinach	*Ipomoea aquatica*
African oxygen weed	*Lagarosiphon major*
Ambulia	*Limnophila sessiliflora*
Melaleuca	*Melaleuca quinquenervia*
Arrow-leaved monochoria	*Monochoria hastata*
Monochoria	*Monochoria vaginalis*
Arrowhead	*Sagittaria sagittifolia*
Giant salvinia	*Salvinia auriculata*
	S. biloba
	S. herzogii
	S. molesta
Exotic bur reed	*Sparganium erectum*
Water-aloe	*Stratiotes aloides*

South Carolina Noxious Weed List

Water hyacinth	*Eichhornia crassipes*
Alligatorweed	*Alternanthera philoxeroides*
Water chestnut	*Trapa natans*
Brazilian elodea	*Egeria densa*
Hydrilla	*Hydrilla verticillata*
Water primrose	*Ludwigia hexapetala*
Eurasian watermilfoil	*Myriophyllum spicatum*
Slender naiad	*Najas minor*
Common reed	*Phragmites australis*

Other species of concern in South Carolina that may be added to the state list of noxious weeds include:

Water lettuce	*Pistia stratiotes*
Purple loosestrife	*Lythrum salicaria*
Hygrophila	*Hygrophila polysperma*
Limnophila	*Limnophila* spp.
Giant Salvinia	*Salvinia* spp.

Additional References

For additional information about aquatic and wetland plants in and around South Carolina, the following references are available:

Beal, Ernest O. 1977. *A Manual of Marsh and Aquatic Vascular Plants of North Carolina with Habitat Data.* North Carolina Agricultural Experiment Station Technical Bulletin No. 247. North Carolina State University. Raleigh.

Godfrey, R.K. and J.W. Wooten. 1979. *Aquatic and Wetland Plants of Southeastern United States: Monocotyledons.* University of Georgia Press. Athens.

Godfrey, R.K. and J.W. Wooten. 1981. *Aquatic and Wetland Plants of Southeastern United States: Dicotyledons.* University of Georgia Press. Athens.

Dressler, R.L., D.W. Hall, K.D. Perkins and N.H. Williams. 1987. *Identification Manual forWetland Species of Florida.* IFAS Publication. University of Florida. Gainesville.

Radford, A.E., H.E. Ahles, and C.R. Bell. 1968. *Manual of the Vascular Flora of the Carolinas.* University of North Carolina Press. Chapel Hill.

Tiner, Ralph W. 1993. *Field Guide to Coastal Wetland Plants of the Southeastern United States.* University of Massachusetts Press. Amherst.

U.S.D.A. Soil Conservation Service. undated. *Southern Wetland Flora. Field Office Guide to Plant Species.* South National Technical Center. Fort Worth, Texas.

Acknowledgments

Many individuals have assisted in the pursuit of knowledge, specimens, and photographs of the aquatic plants of South Carolina. Phil Fields, Richard Harvey, Stan Hutto, and Jerry Long all spent many long hours on lakes and rivers providing much field assistance and discussion essential to this manual. We are indebted to them for their input, encouragement and support. The following individuals also provided assistance in the field for which we are grateful: G. F. Aulbach, Bob Barden, Marion Barnes, Richard Bates, Tommy Boozer, Tommy Bowen, N. B. Bryan, Keevin Chavis, David Hancock, Steve Harden, Charlie Horn, John Inabinet, Murray Jeffers, Charlie Johnson, John Knox, Philip Mancusi-Ungaro, Ken Manuel, Glenn Patterson, Jeannie Pickett Eidson, Joe Pinson, David Rembert, Brian Richardson, George Sawyer, David Schiller, Rebecca Sharitz, Larry Smith, Evelyn Snider, Curtis Stockman, George Swearingen, Eric Tuggle, Jim Tuten, Ronnie Villeponteaux and R. K. Williams.

The first edition went through several drafts before it reached its final form. Our appreciation is extended to Wade T. Batson, Jr. who provided the framework for many of the descriptions and to Victoria C. Hollowell who wrote the glossary and prepared the layouts of the first drafts. Dan Rainey and Stan Hutto also assisted us in the preparation of these early drafts. We are grateful to Betty Branham and Lisa Ball who typed many early versions of the manuscript. For the second edition, we appreciate the able assistance of Chris Page and Diane S. Kennedy in formatting and layout.

We are grateful to the following people who reviewed preliminary drafts of the manuscript for the first edition and provided critical comments and suggestions: Richard Bates, Steve Hill, Charlie Horn, John Inabinet, John Knox, Ken Manuel, Larry Mellichamp, Harold Ornes and David Webb. We would like to express our thanks to colleagues, friends and students who have offered suggestions that have resulted in refinements in this second edition.

Most of the photographs were taken by the authors and by Larry D. Smith. Special appreciation is extended to the following individuals who supplied additional photographs essential to this manual: Barry Beasley, figure 129; Alison Fox, 80; Richard Harvey, figure 173; Charlie Horn, figure 193; John Knox, figure 51; Jerry Long, figure 174; Larry Mellichamp, figures 1, 2, 3, 24, 55, 59, 111, 112; Ric Pantonial, figure 142; Jim Preacher, figures 139, 166, 169, 177, 190, 195, 198, 227, 229, 232, 233, 248, 249; South Carolina Wildlife Magazine photographers, figures 143-145; and David Webb, figures 12, 25, 32, 38, 52, 70, 92, 104, 138, 140, 150-152, 187, 202. All photographs in Chapter V were supplied by Larry Dyck who was assisted by Barbara Speziale and Glenn Turner.

I. Submersed Plants

Introduction To Submersed Plants

This chapter includes aquatic plants that have their main vegetative parts completely underwater. However, most of these species produce flowers or inflorescences at or above the surface.

This group includes many introduced or exotic species that grow prolifically and are considered to be problematic. These include hydrilla (*Hydrilla verticillata*), Brazilian elodea (*Egeria densa*), slender pondweed (*Potamogeton pusillus*), curly-leaf pondweed (*Potamogeton crispus*), slender naiad (*Najas guadalupensis*), and Eurasian watermilfoil (*Myriophyllum spicatum*). Some native species can also form large populations.

Some submersed species may appear to be emergent or floating plants. For example, the bracts on the emersed inflorescences of variable-leaf watermilfoil (*Myriophyllum heterophyllum*) are so robust and leafy that they can appear to be emergent plants. Floating bladderwort (*Utricularia inflata*), with its flowering stem held above the surface by inflated branches, may appear to be a floating plant. Also, virtually all submersed species can form detached floating mats that cause them to appear to be floating plants.

The macroscopic algae, *Chara* and *Nitella*, strongly resemble submersed vascular plants; however, they are treated in Chapter V with the other algae. Under certain environmental conditions and growth stages, some aquatic plants that are included in other chapters may be found completely submersed. These species are treated in their appropriate chapters as listed below.

Floating Plants-Chapter II

Variable-leaf pondweed, *Potamogeton diversifolius*

Illinois pondweed, *Potamogeton illinoensis*

Shoreline and Wetland Plants-Chapter III

Lemon bacopa, *Bacopa caroliniana*

Water-starwort, *Callitriche heterophylla*

Water-pod, *Hydrolea quadrivalvis*

Creeping primrose, *Ludwigia palustris*

Bog moss, *Mayaca fluviatilis*

Shade mudflower, *Micranthemum umbrosum*

Bur-reed, *Sparganium americanum*

Grasses, Sedges, and Rushes-Chapter IV

Spikerush, *Eleocharis* spp.

Bulrush, *Scirpus* spp.

Creeping rush, *Juncus repens*

Isoetaceae

Isoetes spp.

Quillworts

Figures 1-3

Quillworts are perennial ferns that arise from flattened corms. The leaves narrow abruptly just above the widened bases, and become elongate and narrowly linear. They are rounded or somewhat 4-angled and divided into 4 hollow longitudinal cavities or chambers.

Quillworts do not form true flowers; instead, spores are formed in special structures called sporangia. One sporangium is formed in the dilated basal portion of each leaf. Sporangia are of two types, microsporangia and megasporangia, which produce microspores and megaspores, respectively. Microspores or "male spores" are extremely minute and usually gray in color. Megaspores or "female spores" are white and larger than the microspores, but are usually no greater than $^{1}/_{50}$ inch in diameter.

There are at least five species of *Isoetes* in South Carolina; however, ongoing taxonomic research may change this number. Identification to species is mainly dependent on microscopic features of the megaspores.

Quillworts grow in a variety of habitats and can be found submersed or emergent depending on the type of habitat and water level. In South Carolina, several species are found growing in wet or temporarily flooded areas or are associated with vernal pools on granite outcrops in the piedmont. Our most common species, *Isoetes engelmannii,* grows in shallow water of ponds, lakes and rivers throughout the state. Its leaves can grow up to 3 feet in length.

Quillworts do not have the appearance of typical ferns and are often referred to as "fern-allies." Superficially, quillworts resemble tufted grasses or may be confused with those species of arrowheads *(Sagittaria)* that form linear, bladeless leaves or phyllodia.

1
Growth aspect of small plants

2
Individual plant in clear, shallow water

3
White, dilated leaf bases containing sporangia

Potamogetonaceae

Potamogeton crispus **L.**
Curly-leaf Pondweed
Figures 4-7

4
Extensive growth

5
Zigzag stem and alternate leaves

6
Axillary turion and leaves with undulate margins

7
Auger-shaped turion

Curly-leaf pondweed is a rhizomatous and stoloniferous perennial that grows in water as deep as 15 feet. The plants are highly branched with zig-zag, strongly flattened and somewhat brittle stems. The upper portions of the plant grow horizontally just beneath the surface of the water. Short branches with fan-like clusters of leaves are produced in the axils of the upper leaves.

All leaves are submersed, alternate, oblong to widely linear, and sessile. The base of the blade weakly clasps the stem and the apex is rounded. The margins are sharply toothed and undulate, and the midrib is prominent. The leaves are usually 2-3 inches long but can range from 1-4 inches. They are usually $^1/_4$-$^1/_2$ inch wide. The upper leaves are often tinged with red.

Flowers produced in the spring and summer are arranged in spikes borne on axillary peduncles. The fruit is small and 1-seeded with a long beak. Studies in other states indicate that the seeds do not germinate.

Vegetative propagules are produced in the spring at the ends of lateral branches, often in large numbers. They have been variously referred to as winter buds, turions, or dormant apices. They are dark green, hard, $^1/_3$-1 inch long and $^1/_2$ inch wide. Germination of the propagules in more northern states occurs in the fall, the new plants overwinter and then grow rapidly in the early spring.

Curly-leaf pondweed is an aggressive aquatic plant introduced from Europe. It was first documented in South Carolina in the coastal plain in 1990 and now occurs in coastal plain and piedmont lakes and rivers. It can form large, dense populations that severely impair recreation and other water uses.

Potamogeton pusillus L.

Slender Pondweed

Figures 8-11

Slender pondweed is a submersed perennial plant that lacks rhizomes; the stems root at the nodes. At least in some parts of its range it forms corm-like winter buds. Floating leaves are absent; the submersed leaves are alternate, linear, 1-2 inches or more long, and entire. The leaves usually have a pair of inconspicuous glands at the base. The stipules are free from the leaf base and are $^{1}/_{3}$-$^{2}/_{3}$ inch long.

The flower stalks arise from the upper leaf axils and support short, headlike spikes. The achenes are about $^{1}/_{10}$ inch long, obovoid, plump, smooth and rounded on the back, and beaked.

Potamogeton pusillus can be confused with *Najas guadalupensis* but is distinguished by its alternate, entire leaves and small spikes of flowers and fruits produced on peduncles in the leaf axils. When *Potamogeton diversifolius* occurs without floating leaves, it can also be mistaken for *P. pusillus*. However, *P. diversifolius* is distinguished by having fused stipules, keeled achenes, and no glands at the base of the leaf.

Other pondweed species that may resemble slender pondweed include sago pondweed, *P. pectinatus*, and leafy pondweed, *P. foliosus*. However, these species are not found frequently in South Carolina. Sago pondweed differs from slender pondweed by having stipules that are $^{3}/_{4}$-2 inches long and attached to the leaf base. The achene of leafy pondweed differs by having a toothed, winged keel.

Slender pondweed grows in water up to 16 feet deep and may form dense beds in shallower water. It is often associated with other submersed aquatics such as *Hydrilla, Egeria densa, Najas minor, N. guadalupensis* and with other pondweed species. It is found in lakes and ponds throughout the state.

8
Dense growth of flowering plants

9
Appearance of a tangled mat

10
Alternate, linear leaves with stipules

11
Flowering spike and peduncle

Najadaceae

Najas guadalupensis (Sprengel) Magnus

Southern Naiad

Figures 12-14, Plate III-D

12
Growth aspect

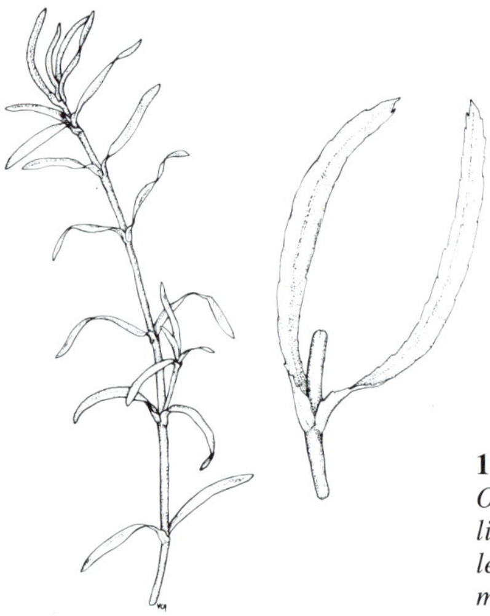

13
Opposite, linear leaves with minute teeth

14
1-seeded fruit in leaf axil

Southern naiad is a submersed, monoecious annual with slender, branched stems that grow several feet in length. The leaves are opposite, narrow, linear, and mostly $^1/_4$ -$^3/_4$ inch long or occasionally, to 1 inch long. The margins are minutely toothed with 50-100 teeth per side; the teeth are evident with magnification. The leaf bases sheathe the stem; each side of the sheath has 5-8 teeth on the margin. The internodes are about as long as the leaves, or longer, reaching $3^1/_2$ inches on the lower stem.

The flowers are unisexual, inconspicuous, and produced in the axils of the leaves. The fruits are 1-seeded, about $^1/_{10}$ inch long, brown and pointed at both ends. They are produced in large numbers throughout the summer and the fall.

Southern naiad can be confused with *Potamogeton pusillus* which differs in having entire, alternate leaves and several achenes in a short peduncled spike.

This species is often found mixed with other submersed plants such as *Najas minor*, *Egeria densa* and *Potamogeton pusillus*. Southern naiad is native to our area and grows in lakes and ponds throughout the state.

Najas minor Allioni
Slender Naiad, Brittle Naiad, Spinyleaf Naiad
Figures 15-18, Plate III-B

Slender naiad is a submersed monoecious annual with profusely branched stems that grow to 4 feet or more in length. The leaves are opposite and linear, and mostly 1-1$^1/_2$ inches long with 7-15 marginal teeth which are usually evident without magnification. The leaves become stiff and recurved with age. The leaf bases sheathe the stem and have 5-8 teeth on each side. The internodes on the main stem often reach 2$^1/_2$ inches in length. However, internodes on the branches, especially near the ends, are very short so the leaves appear to be whorled or in clusters giving the plant a bushy appearance.

The flowers are inconspicuous and are produced in the leaf axils. Fruits are 1-seeded, recurved and about $^1/_8$ inch long; they appear ribbed with magnification. Fruits are produced from late spring through fall.

Reproduction is by seed and by fragmentation. The plants are brittle and readily break into fragments when disturbed. Each fragment is capable of taking root and producing another plant. Slender naiad can be found in pure stands or in association with other submersed species such as *Potamogeton pusillus*, *Najas guadalupensis*, *Hydrilla verticillata* and *Egeria densa*.

Slender naiad is an introduced species from Europe that has rapidly spread throughout the state, growing in lakes and ponds to depths of over 12 feet.

Najas gracillima is a native species of naiad that is similar in appearance to the early season form of slender naiad; however, it is more slender and delicate. It occurs occasionally in mountain and piedmont lakes in South Carolina.

15
Growth aspect

16
Crowded, opposite leaves

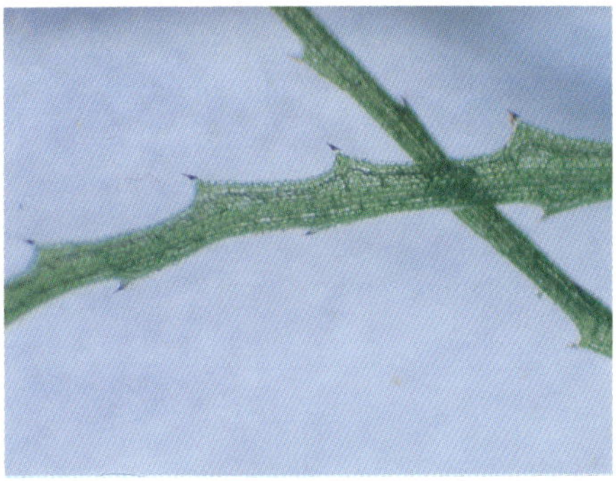

17
Toothed leaf margins

18
1-seeded fruit in leaf axil

Hydrocharitaceae

Egeria densa Planchon

Brazilian Elodea, Egeria

Figures 19-22, Plates III-E, VI-B

19
Growth aspect of extensive population

20
Growth habit of flowering plants

Egeria is a branched, submersed, dioecious perennial that is found rooted in the substrate or free-floating in the water. The stems are mostly $^1/_{15}$-$^1/_6$ inch in diameter and grow to lengths of 6 feet or more. Branching usually occurs where there are two nodes (double nodes) very close together; this is sometimes referred to as a "double internode." Adventitious roots are formed at the branching points. Only 1 or 2 branches are usually formed at each double node and only those fragments containing a double node can develop into new plants.

The leaves are most often found in whorls of 4 but also occur in whorls of 3-6. There may be up to 10 leaves per whorl at the branching points. The leaves are spreading, narrowly oblong or linear, 1-1$^1/_4$ inches long and $^1/_{10}$-$^1/_5$ inch wide. The very finely toothed or serrulate leaf margins are barely evident without magnification.

The flowers are unisexual; only staminate plants have been introduced to the United States thus far. Flowers are produced in sessile spathes in the axils of the leaves. There are 2-4 flowers per spathe which usually mature one at a time. The flowers are borne on pedicels 1-4 inches long which raise the flowers up to the surface of the water. The flowers are showy, having 3 small green reflexed sepals and 3 white spreading petals. The petals are up to $^1/_3$ inch or more long. There are 9 stamens in 3 whorls; the anthers are bright yellow. Flowering can occur throughout the growing season but is most abundant in the spring and early summer.

Brazilian elodea is similar in appearance to *Hydrilla verticillata* and *Elodea canadensis*. However, it can be distinguished by its generally more robust size, rich green color, and smooth texture. Table 1 and Figure 25 provide detailed comparisons of these species.

Vegetative growth can be very rapid in the spring and summer and this species readily spreads by fragmentation. Prior to the introduction of hydrilla, Brazilian elodea was the most problematic aquatic plant in the state. It forms large dense populations that suppress or eliminate other submersed aquatic plants, restrict water withdrawals, and impair recreational water use activities. Thick surface mats may support emergent species, such as water primrose, alligatorweed, and water pennywort. Hydrilla is the only submersed aquatic species in South Carolina that appears to be capable of outcompeting Brazilian elodea. In water bodies where they coexist, hydrilla readily becomes the dominant species.

This introduced plant from South America is now well established throughout the state; it thrives in ponds, lakes, streams and rivers.

21
Whorled toothed leaves; double nodes

22
Showy staminate flower

Elodea canadensis Michaux

Canadian Elodea, Waterweed

Figures 23-24

23
Whorled, serrulate leaves

24
Small staminate flower

Canadian elodea is a submersed, dioecious perennial found either rooted in the substrate or as detached free-floating fragments. The leaves are in whorls of 3's, bright green, narrowly elliptic to elliptic or oblong, acute, and serrulate. They are usually $^1/_4$-$^2/_3$ inch long and about $^1/_{10}$ inch wide.

Small male and female flowers with white petals are produced on separate plants from mid-summer to early fall. Male flowers are about $^1/_3$ inch wide; the female flowers are much smaller, averaging $^1/_6$-$^1/_4$ inch in diameter.

Canadian elodea is similar to Brazilian elodea and hydrilla in general habit but is readily distinguished from them by its considerably smaller size and by having only 3 leaves per whorl. See Table I and Figure 25 for comparisons of these species.

Elodea canadensis is infrequently encountered in South Carolina; it is more common in the northern half of the United States.

Table I. Comparison of Elodea, Egeria and Hydrilla

Elodea canadensis	**Egeria densa**	**Hydrilla verticillata**
Smooth to the touch	Smooth to the touch	Rough to the touch
Serrulate leaf margins barely evident without magnification	Serrulate leaf margins, barely evident without magnification	Sharply toothed leaf margins, usually evident without magnification
Lower midrib of the leaves smooth	Lower midrib of the leaves smooth	Usually 1-12 small, sharp teeth or spines on the lower midrib of the leaf
Staminate flowers $^1/_3$ inch wide, pistillate flowers $^1/_6$ - $^1/_4$ inch wide	Large, showy staminate flowers, $^1/_2$ - $^3/_4$ inch wide	Small, inconspicuous white pistillate flowers, less than $^1/_4$ inch wide
Number of leaves per whorl about the same at branching and non-branching nodes	Number of leaves at branching nodes usually double the number at non-branching nodes	Number of leaves per whorl usually about the same at branching and non-branching nodes; however, they may be double
Leaves in whorls of 3	Leaves in whorls of 3-6	Leaves in whorls of 3-8

Elodea **Egeria** **Hydrilla**

25
Comparison of whorls of ***Elodea, Egeria*** *and* ***Hydrilla*** *leaves*

26
Growth aspect of extensive population

27
Highly branched bushy appearance

28
Dense surface growth covers desirable plants

29
Toothed leaf with typical midrib spines

Hydrilla verticillata (L.f.) Royle

Hydrilla

Figures 26-32, Plates V-D, VI-E, VI-F

Hydrilla is a submersed perennial plant introduced from the Old World. The roots are smooth, wiry, slender, and often orange in color. It is rhizomatous and has several to 15 or more stems from the base; each stem becomes increasingly more branched upward. The nodes are usually single at branching points; 2-8 branches are formed at these branching nodes. Stems are slender, from $^1/_{20}$-$^1/_8$ inch in diameter, and reach lengths of 20 feet or more. Short internodes and numerous branches near the surface of the water result in a bushy appearance.

Hydrilla exhibits very rapid early growth. After reaching the surface of the water, stems grow horizontally producing a dense mat of vegetation at and near the surface. Because of these growth characteristics, hydrilla readily outcompetes other submersed plant species.

Leaves are sessile, oblong and whorled; they are $^1/_3$-$^2/_3$ inch long and $^1/_8$-$^1/_4$ inch wide, and are finely and sharply serrated and spinulose. The midrib on the upper surface can be tinged with red; on the lower surface it usually bears 1-celled, sharp teeth or spines. These teeth are variable in number, even on a single plant, ranging from none to 12 per leaf; where none, at least a few leaves on the plant will show evidence of a few teeth on the midrib. The greater the number of teeth on the midrib, the rougher the plant is to the touch. The number of leaves per whorl is variable, generally at the base there are 3-4; mid-stem 3-6 and upper stems 4-8 per whorl.

Strains of hydrilla can be dioecious with plants producing only pistillate flowers or monoecious with each plant producing both staminate and pistillate flowers. Monoecious plants are usually smaller and less robust than dioecious plants, and often lack spines on the lower midrib of the leaf.

The axillary pistillate flowers of both hydrilla types are small, less than 1/4 inch in diameter, translucent to white, and are produced on the upper branches from summer through fall. Monoecious plants also produce axillary staminate flowers that become free-floating on the water surface. By late fall, the flowers are very conspicuous due to the large number produced.

Rhizomes are scaly, white to reddish-orange to orange-brown, and are formed in the substrate. In late fall and early winter tubers are formed at the ends of the rhizomes of dioecious hydrilla. Monoecious hydrilla begins tuber production in the summer. In fall, turions, or winter buds, are produced in the leaf axils, particularly on stressed plants and fragments. Both turions and tubers give rise to new plants.

Hydrilla can be confused with *Egeria densa* and *Elodea canadensis* but is usually be distinguished from both by its rough texture. Detailed comparisons of these species are presented in Table I. Monoecious hydrilla is distinguished from *Elodea canadensis* by usually having 4 or more leaves per whorl.

Invariably, hydrilla becomes the dominant submersed species shortly after its introduction to a water body. It is the most problematic aquatic species in the state because it spreads rapidly, can severely impair water use activities, has proven to be difficult to control, and usually forms large, dense populations that displace desirable native species. Thick surface mats often support the growth of floating and emergent species, such as water hyacinths and water primrose. Hydrilla has spread rapidly since its introduction to the state in the early 1980's. Currently dioecious hydrilla occurs in reservoirs and rivers from the lowcountry to the upstate. However, monoecious hydrilla may also become established.

30
Whorled leaves with serrate margins

31
Small pistillate flowers

32
Subterranean tubers at ends of rhizomes

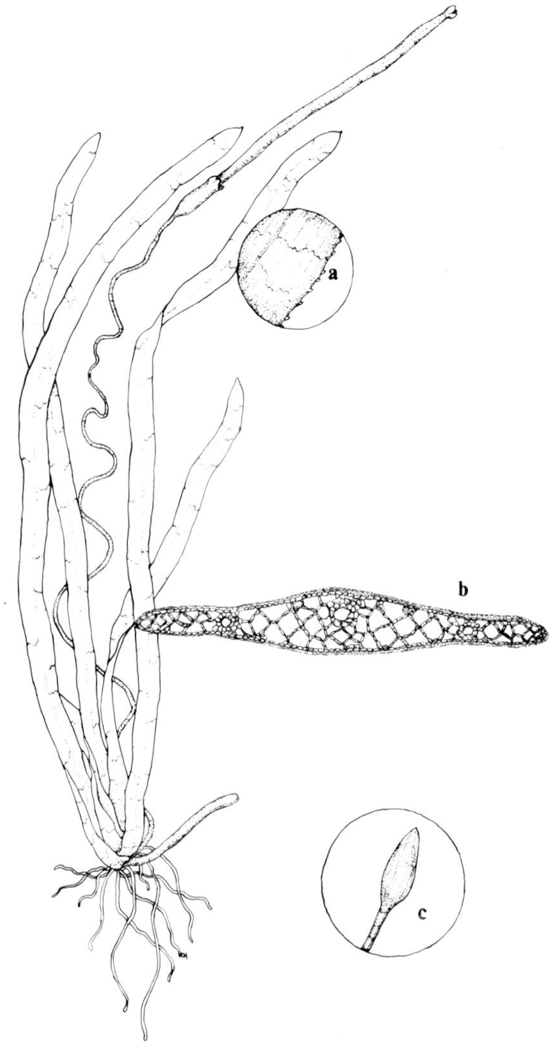

33
Long ribbon-like leaves and pistillate flower;
a) serrate leaf margin;
b) leaf cross-section;
c) staminate spathe

Vallisneria americana Michaux
Tape-Grass, Eel-Grass
Figure 33

Tape-grass is a submersed, dioecious, rooted plant that produces stolons and ribbon-like basal leaves to 7 feet long. The leaves are thin and conspicuously veined with finely serrulate margins; they lack a conspicuous midrib.

Numerous staminate flowers are produced in small spathes at the base of the plant. The spathes open at maturity and release minute flowers that rise to the surface and become free-floating. These flowers consist of 3 sepals, 3 white petals, and 2 stamens. The pistillate flowers are solitary and are borne to the water surface on long, very slender, somewhat coiled peduncles. After fertilization, the fruit is drawn underwater where it ripens into a several-seeded capsule.

Bur-reed, *Sparganium americanum*, can be confused with tape-grass. Bur-reed is typically an emergent plant but can occur in a vegetative, submersed condition, usually when it is growing in flowing water. Bur-reed can be distinguished by its entire leaf margin and by the presence of a midrib. A submersed species of bulrush, *Scirpus etuberculatus*, also resembles tape-grass. It is distinguished by having triangular stems.

Tape-grass flowers in the summer and fall, and it grows in ponds, lakes, and quiet rivers in the coastal plain.

Ceratophyllaceae

Ceratophyllum demersum L.

Coontail, Hornwort

Figures 34-36, Plates II-B, VI-G

Coontail is a rootless, submersed, monoecious, annual plant that grows to lengths of 15 feet or more and can form large colonies. The plants are usually rough to the touch. The stems are either free-floating or are tenuously attached to the substrate by rhizoids. The stems are elongated, branched, and readily fragment. The submersed leaves are in whorls of 5 or more, $^{1}/_{2}$-$^{3}/_{4}$ inch long, and each is dichotomously dissected into linear segments with the segments toothed on one side only. The leaves are crowded toward branch tips.

Flowers are sessile in the leaf axils, solitary, unisexual, minute, and without sepals and petals; they are subtended by several small bracts. The staminate flowers consist only of anthers; pistillate flowers consist of a single ovary producing a single spiny achene. The flowers remain entirely submersed.

Coontail can occur in fairly deep water and often cannot be seen from the surface. It is frequently associated with hydrilla or slender naiad or is found beneath floating mats of vegetation.

Coontail flowers from late summer to fall and grows in ponds and lakes in the coastal plain and occasionally in the piedmont.

34
Growth habit

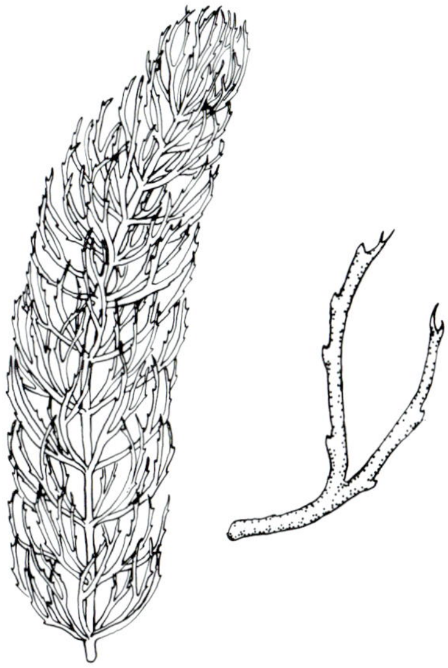

35
Whorled, forked leaves with toothed segments

36
Axillary staminate flower

Cabombaceae

Cabomba caroliniana Gray

Fanwort

Figures 37-39, Plate I-D

37
Growth aspect of flowering plants

38
Opposite, dissected, fan-shaped leaves

39
Flower of the purple form

Fanwort is a submersed annual plant with many flexuous, slightly flattened stems arising from the base; the roots are fibrous and silvery. The leaves are opposite and dichotomously dissected into linear divisions, resulting in a fan-like shape; these leaves are soft and flexuous. The upper stem produces alternate, peltate, floating leaves that are small, firm in texture, green, narrowly elliptic, and about $^{1}/_{2}$-$^{3}/_{4}$ inch long.

The flowers are about $^{1}/_{3}$ inch wide and are produced on long pedicels from the axils of the floating leaves. These flowers are held just above the water surface when fully open. Each flower consists of 3 sepals and 3 petals that are similar in color and shape; however, the petals are spotted with yellow to yellow-orange at the base. There are 6 stamens which have yellow-orange anthers and green filaments and there are usually 3 pistils which are attached at the base. A small cluster of pod-like fruits is formed from each flower. Each fruit contains a few seeds.

The color of the stems, leaves and flowers is variable, ranging from green with white or cream flowers to reddish-purple with pink to light purple flowers. The purple form has occasionally been considered a distinct variety or species.

Fanwort flowers through the summer and grows in lakes, ponds and rivers in the coastal plain. It is occasionally found in the lower piedmont.

Podostemonaceae

Podostemon ceratophyllum Michaux

River-weed

Figures 40-42

River-weed is a unique, submersed, alga-like plant found in flowing waters. The plants are firmly attached to rocks and other hard substrates by rhizoids and by small disk-like holdfasts called haptera. Depending on environmental conditions the plants can vary from dark green to reddish-green in color. These herbaceous plants range from $1^1/_2$ inches to over 8 inches in length. The stems are usually rounded and bear alternate leaves. The leaves are 2-ranked on the stem and are usually divided into linear or filiform segments.

The small perfect flowers are solitary in the axils of the leaves and open only when they are exposed to air when water levels drop. The fruit is a capsule that releases its seeds only when it is dried out.

River-weed can completely cover submersed rocks with a dense, tightly-knit mat or it can be elongate in rapidly flowing water. This is the only species of *Podostemon* in the United States and Canada. It is typically found in rocky shoal areas of piedmont and mountain streams in South Carolina.

40
Typical habitat

41
Dense growth on exposed rock

42
Alga-like appearance

43
Growth aspect of submersed plants

44
Growth aspect of emersed flowering stems

45
Small flowers in axils of whorled bracts

Haloragaceae

Myriophyllum heterophyllum Michaux

Variable-leaf Watermilfoil, Two-leaf Watermilfoil

Figures 43-45, Plate II-C

This perennial, submersed plant is found rooted in up to 6 feet of water. The plants are thick and cylindric and have robust stems that are usually dark red, branched, and up to $^1/_3$ inch in diameter. The pinnately dissected leaves are in whorls of 4-6, grow up to 2 inches long, and have numerous capillary divisions.

The flowering stems are emersed and are 6-12 inches or more in length. The flowers are inconspicuous and occur in whorls around the stem. They are subtended by conspicuous, leafy bracts which are lanceolate to ovate and usually serrate. Because of the size of the flowering stems and bracteal leaves, when variable-leaf watermilfoil is in flower, it has the appearance of an emergent plant as shown in Figure 44.

When water levels are low in late summer and fall, juvenile aerial shoots can be produced along exposed, moist margins. The leaves of these shoots strongly resemble the inflorescence bracts of *M. pinnatum,* the cut-leaf watermilfoil. However, no flowers are formed in the axils of these aerial shoots of variable-leaf watermilfoil.

Variable-leaf watermilfoil can be found in flower from late April to August and grows in lakes, ponds, swamps, and rivers throughout the coastal plain and occasionally in the lower piedmont.

Myriophyllum laxum Shuttleworth ex Chapman
Sandhills Watermilfoil, Piedmont Watermilfoil

Figures 46-47

This native watermilfoil is vegetatively similar to variable-leaf watermilfoil. However, the submersed, vegetative stems are very slender, less than $^{1}/_{10}$ inch in diameter, and usually red in color. The stems typically have scattered, raised black spots and they produce many adventitious roots. The leaves have fewer capillary divisions than those of the variable-leaf watermilfoil, usually 4-8 pairs.

The inflorescence is emersed and slender, from 1-7 inches in length. The upper inflorescence bracts are usually no longer than the flowers that they subtend and are narrowly lanceolate and entire; they are frequently alternate on the flowering stem rather than whorled, as in the variable-leaf watermilfoil. Flowers can be found from May to October.

Sandhills watermilfoil was once considered rather rare in South Carolina. However, it is now known to be fairly common, especially in sandhills millponds, and it can form large populations that impair recreational water use. *Utricularia olivacea*, a very small and rare bladderwort, is often found entangled in the submersed leaves of the sandhills watermilfoil. Another common associate is *Utricularia floridana*, the Florida bladderwort.

Although sometimes called the piedmont watermilfoil, this species does not occur in the piedmont of South Carolina. It is found in ponds in the sandhills and inner coastal plain.

46
Reddish-brown aspect of submersed plants

47
Emersed flowering stems with staminate (upper) and pistillate (lower) flowers

48
Flowering plants in a water-filled ditch

49
Small flowers in axils of whorled pinnate bracts

Myriophyllum pinnatum (Walter) B.S.P.

Cut-leaf Watermilfoil

Figures 48-49

The cut-leaf watermilfoil is similar to the sandhills watermilfoil because it has slender stems and less than 10 pairs of capillary divisions per leaf. However, the bracts of the emersed flowering stems are very different. These bracts are usually whorled, are longer than the flowers, and are pinnately divided.

At certain times of the year, particularly late summer and fall, it is possible to confuse *Myriophyllum pinnatum* with variable-leaf watermilfoil, *M. heterophyllum.* When water levels recede or when *M. heterophyllum* is stranded on moist soil, it produces juvenile aerial shoots that strongly resemble the emersed flowering stems of *M. pinnatum.* However, no flowers are formed in the axils of the juvenile leaves of *M. heterophyllum.*

Cut-leaf watermilfoil flowers during the summer and early fall. It is found throughout the coastal plain in shallow water or stranded on mud along the margins of lakes and ponds.

Myriophyllum spicatum L.
Eurasian Watermilfoil
Figures 50-52

Eurasian watermilfoil is an introduced species that is distinctly different from our native species of watermilfoil. It has long flexuous stems with the leaves concentrated toward the upper ends, especially in turbid water. Actively growing portions of the plants often have a red coloration. The leaves are usually bright to dark green, in whorls of 4, up to 1 inch in length, and have at least 12 pairs of capillary segments. These segments are fairly stiff, compared to those of the other watermilfoils, so the leaves have a very regular, untangled, feather-like appearance.

The inflorescence bears a resemblance to that of the sandhills watermilfoil, *M. laxum,* in that the flowering stem is very slender and the bracts subtending the flowers are inconspicuous. Flowering generally occurs in the summer to early fall; however, plants may be found in flower throughout the summer.

Eurasian watermilfoil was first documented in South Carolina in 1972; since then, it has been found in a few other sites in the state. It grows in lakes, ponds and rivers and is quite tolerant of fluctuating water levels and strong flows. Because much of the vegetative growth is concentrated near the surface, it has the capability to shade out other species of aquatic plants. Eurasian watermilfoil is considered extremely problematic in other parts of the country and the world. Because it readily spreads by fragmentation, its potential for becoming a serious threat to water bodies in South Carolina is great.

50
Growth aspect of submersed plants

51
Whorled feather-like submersed leaves

52
Flowering stem with staminate (upper) and pistillate (lower) flowers

Lentibulariaceae

Utricularia spp.

Bladderworts

Figures 53-60, Plate V-I

53
Utricularia floridana *in flower*

54
Growth habit of ***Utricularia floridana***

55
Distinctive purple flowers of ***Utricularia purpurea***

56
Showy yellow flowers of ***Utricularia juncea***

Bladderworts are carnivorous, aquatic and wetland plants that range in size from the diminutive and rare *Utricularia olivacea* to the robust *U. floridana.* Flowers, produced on an emergent scape, are usually yellow, but are white in *U. olivacea* and purple in *U. purpurea.* Most bladderworts are aquatic plants, but several are terrestrial plants that grow on moist soil or along margins of water.

Bladderworts do not have true roots; the plants are composed of stems and branches which may be leaflike. Bladders, or traps, are present on all or a portion of the stems and branches. These bladders capture small invertebrates which are then digested by enzymes, thus providing the plants with nitrogen and other essential nutrients.

There are 10 or 11 species of bladderworts that are known to occur in South Carolina. Identification to species is usually dependent on details of the flowering scape and flower structure. Bladderworts are found throughout the state; however, most species are restricted to the sandhills and coastal plain.

Fibrous bladderwort, *Utricularia fibrosa,* and Florida bladderwort, *U. floridana*, are characterized by having typical bladder-bearing branches and cylindrical, foxtail-like branches that are bladderless. *U. fibrosa* is more common in South Carolina, occurring thoughout the coastal plain whereas *U. floridana* appears to be restricted to the fall line sandhills.

Two-flowered bladderwort, *Utricularia biflora*, is found throughout the state in shallow water of ponds, lakes, marshes, ditches and bogs. It has very slender, creeping or floating stems and branches that usually form mats. The leaf-like branches are composed of alternate filiform segments and bear numerous bladders. The flower stems are erect, emersed and up to 4 inches high. They bear 1-3 yellow flowers on short pedicels The calyx is 2-lobed; the corolla is very irregular, 2-lipped, spurred, showy, and $^{1}/_{4}$-$^{1}/_{3}$ inch wide. The capsules are small and few-seeded.

In contrast to *Utricularia biflora* is the floating bladderwort, *U. inflata*. This robust species has elongate stems that bear two types of leaf-like branches: submersed, vegetative branches and floating, inflated branches. The submersed, vegetative branches are alternate, bushy, and bladder-bearing. They are dichotomously branched with many of the segments on a particular branch formed in one plane. As a result, they appear to form pentagons, a distinctive field identification feature.

The inflorescence of floating bladderwort is supported by a floating whorl of 5-10 spongy inflated branches bearing finely dissected segments; these branches are 2-3 inches long. The flowering scape is emersed, up to 6 inches in length and bears up to 18 flowers. The flowers are showy, yellow, and spurred, up to $^{3}/_{4}$ inch wide. The capsules are small and borne on recurved stalks.

Floating bladderwort flowers abundantly in the spring and then sporadically throughout the growing season. It is usually found in the submersed, vegetative stage or occasionally with remnants of the floating branches. It grows in lakes, swamps ditches and bogs in the coastal plain.

57
Flowering plants of ***Utricularia biflora*** *in shallow, marshy area*

58
Showy yellow flower of ***Utricularia biflora***

59
Floating whorled branches and flowers of ***Utricularia inflata***

60
Bladders of ***Utricularia inflata***

II. Floating Plants

Introduction to Floating Plants

Plants included in this chapter are those in which the leaves, or the whole plant, are usually found floating on the surface. Some of the species are rooted in the substrate, while others are completely free-floating. To assist in their identification, the species in this chapter are listed below based on this distinction.

Free-floating Plants
Giant salvinia, *Salvinia* spp.
Mosquito fern, *Azolla caroliniana*
Water lettuce, *Pistia stratiotes*
Duckweeds, *Lemna* spp.
Giant duckweeds, *Spirodela* spp.
Water meal, *Wolffia* spp.
Bogmat, *Wolffiella floridana*
Water hyacinth, *Eichhornia crassipes*
Frog's bit, *Limnobium spongia*

Rooted Floating Plants
Pondweeds, *Potamogeton* spp.
Watershield, *Brasenia schreberi*
Yellow lotus, *Nelumbo lutea*
Spatterdocks, *Nuphar* spp.
Water-lily, *Nymphaea odorata*
Floating-hearts, *Nymphoides* spp.
Frog's bit, *Limnobium spongia*

Leaves of rooted floating plants usually lie flat on the water surface. However, leaves of spatterdock and yellow lotus can extend above the water surface as they mature, making them appear more like emergent plants.

Some of the species treated in this chapter may be found entirely in the submersed condition depending on the growth stage, season, or certain environmental conditions. These include variable-leaf pondweed (*Potamogeton diversifolius*), Illinois pondweed (*Potamogeton illinoensis*) and occasionally, spatterdock (*Nuphar luteum*).

Species included in other chapters that may occasionally appear as floating plants under certain conditions include:

Submersed Plants - Chapter I
Floating bladderwort, *Utricularia inflata*

Shoreline and Wetland Plants - Chapter III
Water pennywort, *Hydrocotyle ranunculoides*
Smartweeds, *Polygonum* spp.
Parrot-feather, *Myriophyllum aquaticum*
Water primrose, *Ludwigia hexapetala*

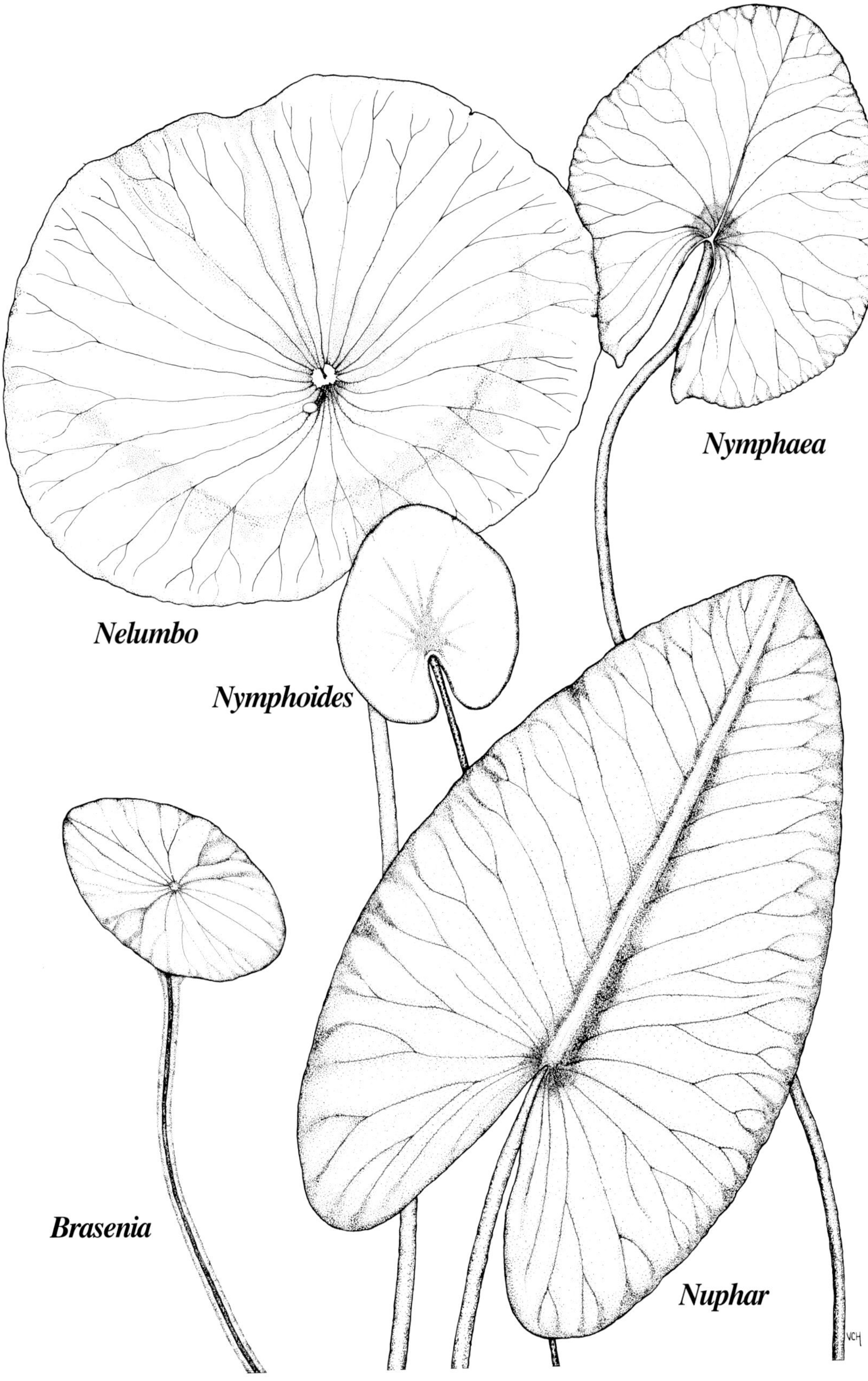

61
Leaf shapes and venation patterns in common species of plants with floating leaves

Salviniaceae

Salvinia molesta Mitchell

Giant Salvinia

Figures 62-64

62
Thick floating mat in small pond

63
Opposite floating leaves with distinct midrib

64
Dense hairs on leaf surface

Giant salvinia is an introduced aquatic fern that floats on the surface of the water and has been known to form extensive mats up to 3 feet thick. Three distinct growth stages for giant salvinia have been described: survival, colonizing and mat-forming stages.

The most commonly encountered stage, the colonizing stage, is characterized by branched stems and floating leaves in whorls of 3. There are 2 opposite leaves that lie flat on the surface and a third highly modified, submersed leaf that is divided into segments that look like long feathery roots. The floating leaves are suborbicular, 1/2-1 inch long and wide, usually slightly wider than long. These leaves are slightly notched at the apex and have one central midrib-like structure. The blades are light to medium green and are on very short petioles. The lower surface has long appressed brownish hairs. The upper surface is covered with dense, stiff white hairs arranged in distinct rows. Each hair is branched into 4 segments which then fuse at their tips giving them an "egg-beater" or "bird cage" appearance.

The survival stage occurs during periods of environmental stress and is characterized by smaller leaves. The mat-forming stage is characterized by larger leaves that are folded upward and crowded on the stem. The color varies from dark to olive green or brownish yellow. Mat forming plants may produce small, brown, ball-like sporangia on the root-like submersed leaves. Their linear alternating arrangement is characteristic of this species.

Giant salvinia has been found in scattered localities in the United States, including South Carolina. It is considered extremely noxious.

Azollaceae

Azolla caroliniana Willd.

Carolina Water Fern,
Mosquito Fern

Figures 65-67, Plate II-G

Mosquito fern is a true aquatic fern that is small, branched, and floats on the surface. The small overlapping leaves vary in color from red, reddish-purple, reddish-green to green. The stems, are branched and covered by the overlapping leaves. The roots, when present, are slender and unbranched. The leaves are arranged alternately in two rows and are two-lobed; the lower lobe is under water. Microspores and megaspores are produced in different spore cases at the base of the lower lobe. Bluegreen algae are likely to be present in pockets of the leaves.

Individual plants are smaller than a dime; however, large numbers often occur and from a distance present a red, reddish-green or green color on the surface of the water. The color and the velvety texture are in strong contrast to the shiny green of the duckweeds with which *Azolla* is often associated. The reddish coloration often seen in *Azolla* has been attributed to stress from low nutrient conditions or high temperatures.

Mosquito fern is found in swamps, lakes, ponds and ditches throughout the coastal plain and, sporadically, in the piedmont.

65
*Red **Azolla** mixed with other plants*

66
Reddish-green velvety appearance

67
Overlapping alternate leaves

Potamogetonaceae

Potamogeton diversifolius Raf.

Variable-leaf Pondweed

Figures 68-70

68
Small floating leaves

69
Linear submersed leaves differ from floating leaves; side-view of achene

70
Stipule fused to leaf base

This is a somewhat delicate branched perennial that grows from slender rhizomes. The plants usually have both floating and submersed leaves present. The floating leaves are subopposite, petiolate, elliptic to oval, up to 2 inches long (usually 1/2-1 inch) and mostly 7-11 nerved. The submersed leaves are alternate, thin and linear, 1/2-1 or occasionally 2 1/2 inches long; they are 3-nerved, but only the central nerve or midvein is conspicuous. The stipules of submersed leaves are up to 1/3 inch long. The lower portion of the stipule is adnate or fused to the leaf base and forms a sheath around the stem.

The flowering and fruiting spikes are produced in the axils of the leaves; those from the floating leaves are on stout peduncles and are held just above the water surface. The spikes and peduncles from the axils of the submersed leaves are shorter than those from the floating leaves. The fruit is a suborbicular flattened achene with a prominent bumpy dorsal keel. The achenes are 1/16 inch or less wide and, from the side, they often appear ear-shaped. Flowers and fruits can be found from spring to fall.

This species is distinguished from other pondweeds by the small size of the floating leaves. Populations are occasionally found with no floating leaves present; when this occurs, it can be confused with *Potamogeton pusillus*. The plump beaked achene of *P. pusillus* and the obvious free stipules distinguish it from *P. diversifolius*.

Variable-leaf pondweed is very common in South Carolina and grows in shallow water of ponds, streams and lakes throughout the state.

Potamogeton nodosus Poiret

Longleaf Pondweed, American Pondweed

Figures 71-73, Plate VI-C

These plants are perennials from rhizomes; the stems are elongate, growing to 6 feet long, and they root at the lower nodes. Both floating and submersed leaves are usually present. The submersed leaves, which are frequently found in various stages of deterioration, are thin and flaccid, linear-lanceolate, up to 8 inches long and 1 1/4 inches wide, on petioles to 6 inches long. The floating leaves are smooth and often shiny on the upper surface, elliptic, 3 1/2-5 inches long and to 1 1/2 inches wide. The midnerve is very prominent beneath and the lateral veins are prominent above. The petioles of floating leaves are smooth and long, mostly to 6 inches long; the stipules are up to one-half the length of the petiole.

The flowers are produced from May through October on emersed thick cylindrical spikes to 1/3 inch wide and 1-2 1/2 inches long. These spikes are on long stout peduncles produced in the axils of the floating leaves. The peduncles are usually thicker than the stems and are 1 1/2-4 or more inches long. The achenes are obovate, smooth, about 1/6 inch long and prominently keeled.

Longleaf pondweed is readily distinguished from *Potamogeton pulcher* in that the base of the leaf is usually rounded to cuneate, and the stems and petioles are smooth. It occurs sporadically throughout the coastal plain but is usually abundant where it does occur. See also *Potamogeton illinoensis* on page 32.

71
Growth aspect

72
Floating leaves and submersed leaf; side view of achene

73
Flowering and fruiting spikes

74
Growth aspect of ***Potamogeton pulcher***

Potamogeton pulcher Tuckerman

Spotted Pondweed

Figures 74-75

Spotted pondweed is an unbranched perennial aquatic arising from rhizomes. The stems and petioles are conspicuously black-spotted. The leaves are of two types, floating and submersed. The submersed leaves are variable but similar in many aspects to those of the longleaf pondweed except the margins are undulate. The floating leaves are ovate to elliptic, 2-3 1/2 inches long and 1-2 inches wide. The petioles range from 2-6 inches or more in length. The base of the leaf is typically cordate but can be somewhat rounded. The stipules of the floating leaves are persistent and up to 2 inches long.

Spotted pondweed typically grows in ponds in the coastal plain but also occurs sporadically in the piedmont and mountains.

75
Potamogeton pulcher *- floating leaves and submersed leaf; side view of achene*

Potamogeton illinoensis Morong

Illinois Pondweed

Figure 76

Illinois pondweed usually occurs in the submersed condition; but, in South Carolina, it is often found with both floating and submersed leaves. The floating leaves resemble those of *Potamogeton nodosus*, but they are usually obtuse at the apex. The submersed leaves resemble those of *P. pulcher* in that they are undulate to somewhat crispate; they can reach up to 1 3/4 inch in width. However, the submersed leaves lack the toothed margins found in *P. crispus.*

Illinois pondweed occurs in lakes in the lower piedmont and the coastal plain.

76
Floating leaves of ***Potamogeton illinoensis***

Hydrocharitaceae

Limnobium spongia (Bosc.) Steudel

Frog's Bit

Figures 77-79, Plates V-B, VI-D

These are dimorphic, monoecious and perennial plants that spread by seeds, stolons and by floating rosettes. The young or juvenile form is usually a small, free-floating rosette, with 4 to many leaves on short petioles. These leaves are heart-shaped, 1/2-1 1/4 inches long and wide, and the margins are entire. The plants float by means of a mass of inflated reddish-purple spongy cells on the underside of the leaf. The rosettes give rise to the emersed, mature form.

Mature plants reach a height of 1 foot above the water surface and often grow in large, tangled mats in shallow water. The emersed leaves are in rosettes, with petioles from 2 1/2-8 inches long; these petioles are spongy and often thickened. The blades are entire and widely ovate with rounded, truncate or slightly cordate bases. These leaves have up to 7 prominent palmate veins from the base, interconnected by many small cross-veins. The plants are connected by stolons that are up to 1/4 inch in diameter. Stolons root at the nodes and produce new plants; they also remain alive just below the water surface during the winter. Mature plants can be rooted in shallow water, or they can grow in floating mats.

Staminate and pistillate flowers are formed in mid to late summer. They are borne on short, thick peduncles arising from the base of the plant and are hidden by the leaves. The fruit is a many-seeded berry.

Frog's bit grows in lakes, marshes, swamps and bogs in the coastal plain. The mature form of frog's bit often resembles water hyacinth.

77
Juvenile floating rosette stage

78
Floating mat of mature plants

79
Pistillate flower and fruits

Araceae

Pistia stratiotes L.

Water Lettuce

Figures 80-81

Water lettuce is a stoloniferous perennial plant that forms large colonies of floating plants. The leaves are formed in rosettes, are densely pubescent, grayish to light green in color and up to 6 inches long. The leaf shape is ovate to obovate with a usually truncate apex. Its method of reproduction is vegetative, producing new plants on stolons.

Water lettuce is an occasional introduction to South Carolina, capable of producing large numbers of plants during the growing season and impacting water use. However, it is intolerant of freezing temperatures and normally does not survive the winter season in South Carolina.

80a
Overview of large population

80b
Dense growth of floating plants

81
Individual rosettes with truncate leaves and parallel veins

Lemnaceae

Duckweed Family

Figures 82-84

This family is comprised of small floating aquatic plants that are not differentiated into stems and leaves but consist of a more or less flattened leaf-like structure variously referred to as a frond or thallus.

Duckweeds reproduce asexually both by the production of turions, which can overwinter, and by producing new plants by budding. Buds are produced in pouches and frequently do not separate from the parent plant, thereby resulting in clusters of plants. Duckweeds also reproduce sexually; the plants are monoecious, each frond bears both staminate and pistillate flowers which are minute and, therefore, rarely seen.

The four genera occurring in South Carolina can be distinguished vegetatively without magnification. Useful features include size and color, as well as the number and color of the roots. *Lemna* is characterized by pale green fronds bearing one root; *Spirodela* is usually dark green above and often reddish-purple below and has two or more roots per frond. *Wolffia* and *Wolffiella* bear no roots; *Wolffia* is almost microscopic in size and *Wolffiella* is sickle-shaped, usually occurring in clusters.

Duckweeds are ubiquitous; usually several species grow together, often occurring mixed with *Azolla*.

82
Typical growth of duckweeds in a quiet cove

83
Dense surface coverage

84
Duckweeds and allies often colonize protected areas created by emergent species

Lemna spp.
Common Duckweeds
Figures 85-86, Plate III-A

85
Light green ***Lemna*** *mixed with other plants*

86
Reproduction of ***Lemna*** *by budding*

87
Triangular ***Ricciocarpus*** *mixed with duckweeds*

88
Deeply cleft Y-shaped thalli of ***Ricciocarpus***

Common duckweed fronds are obovate to elliptic, $^{1}/_{10}$-$^{1}/_{8}$ inch long and less than $^{1}/_{10}$ inch wide. They are pale green on both surfaces and have a somewhat keeled, upper surface that bears a few prominent papillae. Each frond has 1 root on the lower surface and 2 pouches along the margins from which new individuals arise. One or 2 seeds may be produced which are released when the parent plant dies.

There are several species of common duckweeds in South Carolina. They are distinguished by the number of nerves per frond and the color of the lower surface. Duckweeds are found throughout the state but are most commonly encountered on the coastal plain.

A small, floating aquatic liverwort, *Ricciocarpus natans*, can be found intermixed with duckweeds, particularly in the cooler months from fall through spring. These plants occur as somewhat fleshy, triangular, light green segments that appear to be paired (Figures 87-88). The undersurface is covered with dark-colored, thin scales that appear like a dense mass of roots.

Spirodela spp.
Giant Duckweeds
Figures 89-90

Spirodela is distinguished from the other genera of duckweeds by the presence of 2 or more roots on each frond. Giant duckweeds are usually dark green on the upper surface and reddish-purple or green on the lower surface. They often occur in clusters of 2 or more fronds joined together.

The fronds of *Spirodela polyrhiza* are $^{1}/_{8}$-$^{1}/_{3}$ inch long and wide, and are broadly obovate or orbicular. The upper surface usually has an off-center reddish-purple dot from which several prominent veins emanate. The lower surface is reddish-purple or green with 5 or more roots.

Fronds of *Spirodela punctata* are usually solid green on the upper surface or the margin is tinged reddish-purple. A median ridge is usually present on the upper surface and 2-4 roots are produced from the lower surface. The fronds are narrowly ovate to oblong and are $^{1}/_{8}$-$^{1}/_{4}$ inch long and $^{1}/_{8}$-$^{1}/_{6}$ inch wide.

The two species of giant duckweeds often grow together and can be found associated with *Lemna* spp., *Wolffia*, and *Azolla*. Both species of *Spirodela* occur in lakes, ponds, marshes, swamps, and bogs throughout the coastal plain and piedmont of South Carolina.

89
Spirodela polyrhiza *with distinctive red dot*

90
Spirodela punctata *with median ridge of papillae*

91
Wolffia *mixed with* ***Spirodela punctata***

92
Extremely small individual plants of ***Wolffia***

Wolffia spp.

Water Meal

Figures 91-92, Plate IV-B

This smallest and simplest of flowering plants is rootless, floats with only its dorsal surface breaking the water and appears as little green pinheads. The plant body has a rather asymmetrically elliptic outline; it is hemispheric below and somewhat flat on top with a single papilla. The entire plant is punctate.

Plants usually proliferate from a pouch at one end of the long axis. Newly formed plants break away from the parent plant as they mature. When flowers are produced, they erupt through the upper surface as a single stamen and a single pistil, and only 1 seed is produced. Large numbers of this plant sometimes occur.

The most common species of water meal in South Carolina is *Wolffia papulifera.* This species grows on the surface of the water of lakes and ponds throughout the state.

93
Cluster of fronds of ***Wolffiella***

Wolffiella floridana

(J.D. Sm.) Thomps.

Bogmat, Mud-midget

Figure 93

Bearing little resemblance to other duckweeds, the green fronds of bogmat are narrowly linear, elongated, slightly curved and $^1/_5$ -$^1/_3$ inch long. Typically, several fronds are clustered together in a group. Usually found with other duckweeds, bogmat is easily overlooked. It occurs sporadically throughout the state in swamps, lakes and ponds.

Pontederiaceae

Eichhornia crassipes

(Martius) Solms-Laubach

Floating Water Hyacinth

Figures 94-96

Water hyacinths are introduced free-floating plants that range from a couple of inches to over three feet in height. The tall form is often called the "bull hyacinth." The leathery leaf blades are elliptic, ovate or suborbicular with parallel veins. These blades range from 1/2 inch to over 8 inches in length. The petioles are commonly inflated and spongy; in smaller plants, they are often quite bulbous. The roots are fibrous and feathery, usually purplish to almost black in color.

The inflorescence consists of a spike of 5-20 very showy light purple to blue flowers; the upper petal has a prominent yellow spot surrounded by purple. The fruit is a many-seeded capsule; however, mature fruits have not been found in South Carolina.

These plants reproduce very rapidly by vegetative runners or stolons and form large mats that can completely obstruct waterways. They are closely related to pickerelweed, *Pontederia cordata*. Water hyacinths are native to Brazil and were first introduced to the United States in the mid-nineteenth century.

Floating water hyacinths have been sporadically found throughout the state. However, they are known to cause problems only in the coastal plain, presumably due to the milder winter conditions in that part of the state.

94
Floating mat of flowering plants

95
Young plants with bulbous petioles and feathery roots

96
Showy purple flowers

Cabombaceae

Brasenia schreberi Gmelin

Watershield, Dollar Bonnet

Figures 97-99, Plates III-C, IV-D

97
Elliptic peltate leaves

98
Red-purple lower leaf surface and petiole with jelly-like coating

99
Purple flower with numerous stamens

Watershield is an herbaceous perennial species that grows from slender rhizomes in the substrate; the roots are black and fibrous. All submersed portions of the plant are covered with a mucilaginous or jelly-like substance. The stems are purple and branched, reaching up to 6 feet in length. The leaves are alternate, floating, peltate and elliptic. They are green on the upper surface and purple to reddish-purple beneath and have entire margins. The leaves are up to $4^1/_2$ inches long and $2^1/_2$ inches wide. The petioles are 4-12 inches long.

The flowers are emersed, borne on stout stalks arising from the upper leaf axils; they are purple and $^1/_2$-$^3/_4$ inch long. The 3-4 sepals and petals are similar in appearance. The stamens are numerous (12-36) and contribute to the showiness of the flowers. Flowering usually occurs from May to August. Several dry indehiscent fruits are formed by each flower; each fruit contains 1 or 2 seeds. Fruits mature from late May through September. The seeds and the vegetative parts of watershield are eaten by waterfowl.

Watershield typically grows in shallow water up to about 5 feet deep. It can occur in fairly pure stands of up to 10 acres and is often associated with submersed plants such as fanwort, bladderwort, and coontail. It is frequently found in association with floating and emersed plants such as fragrant water-lily or water primrose. Watershield occurs throughout South Carolina but is most commonly found in the coastal plain.

Nelumbonaceae

Nelumbo lutea (Willd.) Pers.

Yellow Lotus, American Lotus

Figures 100-102, Plate II-H

This is a stout perennial herbaceous plant with leaves and flowers arising directly from a spongy, elongate rhizome that is embedded in the substrate. All parts of the plant contain a milky latex. The leaves are circular, entire, peltate and 1-3 feet wide. Early in the season and/or when they are young, the leaves are flat and float on the water surface. As the petioles grow, they raise the blades several feet above the water surface. The leaf edges then become undulate and the blades assume a shallow funnel-like shape that is similar to an inside-out umbrella. The leaves are bluish-green to grayish-green in color due to a thick deposit of wax on the upper surface.

The flowers are solitary, pale to golden yellow, very showy, fragrant and 4-10 inches wide when fully open. The sepals and petals are similar, usually numbering 20 or more. The flowers are supported several feet above the water surface on long stalks. The fruits are hard nut-like seeds embedded in pits on the upper surface of an obconic receptacle. The mature receptacle is 2-6 inches wide and becomes hard and dark-colored. During the winter, it breaks off and floats upside-down, dropping the fruits as it slowly decomposes.

Yellow lotus flowers and fruits from June through September. Pure stands of several acres are common; or, it may be found in association with other aquatic plants such as alligatorweed, pondweed and water primrose. It typically grows in up to 6 feet of water in lakes, ponds and marshes throughout the coastal plain.

100
Distinctive growth aspect

101
Large peltate leaf, flower and seed pods

102
Receptacle with nut-like fruit and milky latex

Nymphaeaceae

Nuphar luteum ssp. *macrophyllum* (Small) E. O. Beal

Spatterdock, Yellow Pond-lily, Cow-lily

Figures 103-105

103
Floating and emergent leaves with distinctive midrib

104
Emersed yellow flower

105
Fruits with reddish band below stigmatic disk

Spatterdock is a perennial plant that grows from a spongy rhizome up to $1^1/_2$ inches in diameter. The leaves can be submersed, floating or emersed. The submersed leaves are flaccid and bright green; floating and emersed leaves are firm, slightly roughened and dull green. All leaves are entire, cordate, notched at the base, and pinnately veined with a prominent central midrib. They are typically 6-12 inches long and 5-10 inches wide, or sometimes much larger; and they are usually less than twice as long as wide. The petioles are stout, spongy and grow to 3 feet or more in length.

The flowers are 1 inch or more wide, globose, solitary, and produced on long, stout, spongy pedicels at or above the surface of the water. There are 6 petal-like sepals ranging from green to yellow in color. These sepals are thick, stiff and rounded and never open widely. There are many yellow petals that grade into numerous yellow stamens; both are shorter than the ovary and enclosed by the sepals. There are many pistils united together in an ovoid structure that is constricted at the top. This top is a stigmatic disk comprised of fused stigmas that radiate out from the center.

The fruit is 1-2 inches broad, flat-topped and many-seeded. It is green or yellowish and ovoid; occasionally, the constriction at the top is tinged with red. The fruits ripen above or below the water, detaching at maturity and floating. They then sink to the bottom where they can give rise to a new colony of plants.

Spatterdock flowers from late April to October. It grows in lakes, ponds, and swamps throughout South Carolina but occurs predominantly in the coastal plain.

Nuphar luteum ssp. *sagittifolium* (Walter) E. O. Beal

Spatterdock

Figures 106-107

This subspecies of spatterdock differs from the previous subspecies in the shape of the leaves. The leaves are typically three times as long as wide. They are usually up to 12 inches long and 2-4 inches wide. The leaves are also thinner in texture and only floating and submersed leaves are found. The sinus at the leaf base is shorter than in the more common subspecies of spatterdock.

This subspecies is usually found in flowing water; the leaf margins usually appear undulate due to the flowing water. In South Carolina it occurs in the Pee Dee River Basin.

106
Floating leaves with undulate margins

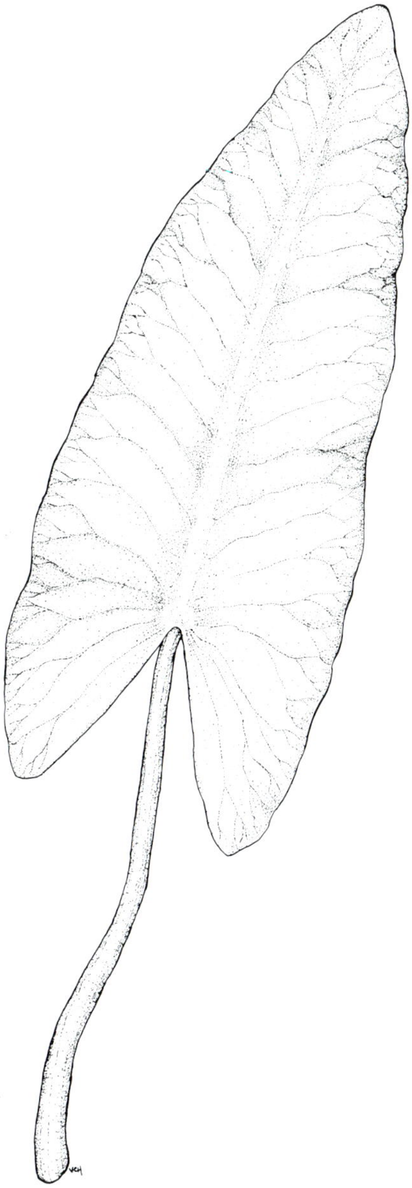

107
Elongate leaf with pinnate venation

Nymphaeaceae

Nymphaea odorata Aiton

Fragrant Water-lily
Figures 108-110

108
Growth aspect

109
Deeply cleft leaves

110
Showy fragrant flowers

Water-lilies are aquatic, perennial plants with leaves and flowers arising directly from stout rhizomes. The leaves are orbicular and floating, usually 6-12 inches wide, green above and usually purplish beneath. The blades are deeply cleft to an almost central stout petiole that is up to 6 feet long. The lobes of the leaf are somewhat pointed, not rounded.

The flowers are showy, solitary, fragrant, floating, and 3-5 inches wide. There are 4 green sepals with scarious or purple-tinged margins. The petals are numerous and white, and there are many yellow stamens. The first day the flowers open, they are functionally female, often trapping pollen-laden insects in a pool of liquid; on the second day, they are functionally male. The fruit matures under water and is surrounded by decaying petals and stamens. Numerous seeds are produced in each fruit.

Water-lilies flower from late spring through fall and grow in ponds, lakes and marshes throughout the state. Cultivated forms of water-lilies have pink, yellow or purple flowers. *Nymphaea mexicana*, known as yellow water-lily or banana water-lily, is an introduced species found primarily in impoundments in the outer coastal plain.

Menyanthaceae

Nymphoides spp.

Floating-hearts

Figures 111-113, Plates I-F, IV-I

Floating-hearts are rhizomatous, perennial plants that have heart-shaped floating leaves. Flowers arise in clusters just below the leaves and are borne on slender pedicels. The flowers consist of 5 sepals and 5 showy white petals that have undulate margins. The fruit is an ellipsoid capsule. Two species occur in South Carolina, *Nymphoides aquatica* (Gmelin) Kuntze and *N. cordata* (Elliott) Fernald.

Nymphoides aquatica is the larger of the two species and is characterized by having red-punctate stems and sepals. The leaves are green above, reddish-purple and rough below and are usually thickened. The leaves are 3-5 inches long. *N. aquatica* is sometimes called 'banana-lily.' This common name refers to the banana-like clusters of tuberous roots that are formed at the junction of the stem and petiole.

Nymphoides cordata, often referred to as 'little floating-heart,' differs from *N. aquatica* in several ways. The stems and sepals are not red-punctate. The leaves are purple and green mottled on the upper surface and the undersurface is purple and smooth. The leaves are smaller, usually no more than $2^1/_2$ inches in length.

Both species of *Nymphoides* flower from spring through early fall and are found in lakes, ponds and streams. *N. aquatica* occurs throughout the coastal plain while *N. cordata* is restricted to the sandhills and inner coastal plain. Occasionally, the two species may occur together in the inner coastal plain.

111
Long petioles and mottled leaves of ***Nymphoides cordata***

112
Green heart-shaped leaves of ***Nympoides aquatica***

113
Nymphoides aquatica *flower*

III. Shoreline and Wetland Plants

Introduction to Shoreline and Wetland Plants

This chapter includes plants having most of the vegetative and flowering parts growing above the surface of the water. Plants in this group are often called "emergent plants" or "shoreline emergents." A comprehensive treatment of shoreline and wetland plants in South Carolina would include close to 2,000 species. Therefore, this chapter presents only the more commonly encountered plants of this large and diverse group.

Although the species in this chapter are usually found as emergent plants, some are occasionally found completely submersed. These species include:

Lemon bacopa, *Bacopa caroliniana*
Water-starwort, *Callitriche heterophylla*
Water-pod, *Hydrolea quadrivalvis*
Creeping primrose, *Ludwigia palustris*
Bog moss, *Mayaca fluviatilis*
Shade mudflower, *Micranthemum umbrosum*
Bur-reed, *Sparganium americanum*

In addition, some species in this chapter form floating mats and therefore may appear to be floating plants. These species include:

Alligatorweed, *Alternanthera philoxeroides*
Water-starwort, *Callitriche heterophylla*
Water pennywort, *Hydrocotyle ranunculoides*
Water primrose, *Ludwigia hexapetala*
Parrot-feather, *Myriophyllum aquaticum*
Smartweeds, *Polygonum* spp.

Many species of grasses, sedges, and rushes are also shoreline and wetland plants. Because they possess unique common characteristics they are treated as a separate group in Chapter IV.

Typhaceae

Typha latifolia L.

Common Cattail

Figures 114-116

Common cattails are robust, erect, perennial plants that grow from creeping rhizomes. The stems are jointless and grow up to 7 feet tall. The leaves are flat, sheathing, green, up to 5 feet long and 1 inch wide.

The flowers are in very dense cylindrical terminal spikes to 6 inches long and lack sepals and petals. The lowermost portion of the spike is dark brown, about 1 inch in diameter, and consists only of stalked pistils; these stalks bear long silky hairs. The upper part of the spike is smaller and consists only of stamens and long hairs. The nutlets are minute and long-stalked.

Typha latifolia flowers in the summer and the fruiting spikes persist through the fall. It grows in shallow water of lakes, ponds, rivers and ditches throughout the state.

Three other species of cattails can be found in the outer coastal plain of South Carolina. *Typha angustifolia*, the narrowleaf cattail, has narrow, dark brown fruiting spikes and narrow leaves. *Typha glauca*, a hybrid between *Typha angustifolia* and *Typha latifolia,* is occasionally found where these two species grow in close proximity. *Typha domingensis*, the southern cattail, is found sporadically in the coastal plain; its fruiting spikes are cinnamon brown in color.

114
Small colony along lake shoreline

115
Long, flat leaves and characteristic "cattail" spike

116
Fruiting spike

Sparganiaceae

Sparganium americanum Nuttall

Bur-reed

Figures 117-119

117
Globular heads of flowers on zigzag stems

118
Long, linear leaves and bur-like fruits

119
Submersed form in flowing water

Bur-reed is a monoecious perennial that grows from rhizomes. The stems are usually robust and grow to 3 feet tall. The leaves are distichous, or 2-ranked, entire, and reach 3 feet in length and 3/4 inch in width. The bases of the leaves sheathe the stem and the blades are linear and spongy-thickened. The lower surface is keeled giving the blade a flattened-triangular appearance in cross section.

The flowers are in globular heads on a terminal zigzag stem. The heads are 1/2-1 inch in diameter; the lower heads are pistillate and the upper are staminate. The fruiting heads are hard and bur-like and contain numerous achenes.

Bur-reed flowers throughout the summer and is quite distinctive when in flower and fruit. It is typically found as an emergent plant in shallow waters of streams, marshes, and ponds throughout South Carolina. Occasionally, it will grow entirely submersed in flowing waters of rivers and streams. When submersed, it can be confused with *Vallisneria* from which it is distinguished by its entire leaves which are spongy-thickened and keeled below.

Alismataceae

Echinodorus cordifolius (L.) Grisebach

Creeping Burhead

Figures 120-122; Plates I-G, III-F, V-A

Creeping burhead is a robust perennial that has clustered, erect, basal leaves. The leaves have long, grooved petioles up to 2 feet or more in length. The blades are widely ovate, rounded at the apex and truncate, cordate or rounded at the base. There are 5-9 prominent palmate veins from the base of the leaf that are interconnected by numerous cross veins. The blades are variable in size, reaching 10 inches in length.

Flowers are produced on a scape that is erect when young but soon arches and becomes decumbent, reaching 3 feet in length. The scape often roots and produces new plants at the nodes. Flowers are in clusters or whorls of 5-15 on long pedicels along the scape. Each flower has 3 showy white petals, 3 sepals, and many stamens and pistils. The pistils mature into a head-like cluster of achenes that are subtended by the sepals.

Creeping burhead flowers throughout the summer and fall, and usually grows in swamps and marshes on moist soil or in shallow water. It occurs in the lower piedmont and the coastal plain.

120
Ovate leaves with cordate bases and arching flowering scape

121
White three-petaled flowers

122
Bur-like fruiting heads of achenes

123
Leaves of ***Sagittaria latifolia*** *with long basal lobes*

124
Inflorescences and narrow elliptic leaves of ***Sagittaria graminea***

125
Inconspicuous linear phyllodial leaves and flower of ***Sagittaria subulata***

126
Details of three-petaled staminate flower of ***Sagittaria***

Sagittaria spp.

Arrowhead, Duck-potato

Figures 123-126; Plates IV-G, V-C

Based on leaf characteristics, the arrowheads appear to be a very diverse group of plants but they share many common characters. They are usually perennial plants that grow from rhizomes and corms. The leaves may vary considerably in shape but are usually in a basal cluster and are petioled if blades are present. Flowers are produced on an erect scape and are usually pedicelled and in whorls of 3. Generally, the lower flowers are pistillate and the upper are staminate. Each flower consists of 3 sepals and 3 showy white petals. The globose fruiting heads consist of numerous flat achenes. Flowers and fruits are produced throughout the summer and fall.

The arrowheads can be split into 3 groups based on general leaf shape. The first group consists of those with the 'typical' elliptic to ovate leaves with long sagittate or hastate basal lobes. In this group is the common arrowhead or duck-potato, *Sagittaria latifolia*. The blades reach lengths of 1 1/2-2 feet and the petioles are up to 3 feet long. Numerous palmate veins originate from the center of the leaf blade then arch and curve following the form of the blade. Common arrowhead can be found in marshes, lakes and low woods throughout the state.

The second group consists of those species that have elliptic, ovate or linear emersed leaves without basal lobes. *Sagittaria graminea* and *S. lancifolia* are typical examples of this group. These species grow in swamps and marshes in the coastal plain; the latter is restricted to the outer coastal plain.

The third group encompasses those species producing only bladeless leaves called phyllodia. *Sagittaria subulata,* which has linear phyllodial leaves, is a good example of this group. This species grows in the outer coastal plain and is restricted to fresh or brackish tidal rivers and marshes. It is usually found submersed or it can be emersed during low tide.

Araceae

Colocasia esculenta (L.) Schott

Wild Taro, Elephant-ear

Figures 127-128

Wild taro is a robust perennial plant that grows from large corms; these rootstocks are commonly referred to as taro-root. Clumps of large leaves arise directly from the corm.

The leaves bear some resemblance to those of *Peltandra* and some species of *Sagittaria*. The blades are peltate, elliptic to ovate with hastate to sagittate rounded lobes. The sinus is rounded, the apex is acuminate and the leaf margins are slightly undulate. The upper surface is medium to dark green and velvety while the lower surface is light to whitish-green. The blades range from less than 10 inches to over 20 inches in length and are 6-12 inches wide. The petioles can reach 3 feet or more in length and the upper portion is often suffused with purple, especially on older and larger leaves. The base of the petioles can reach over 2 inches in width.

Wild taro is a relatively recent introduction to the flora of South Carolina, occurring in shallow water of swamps. Native to Asia, it is naturalized throughout the tropics and subtropics and is widely cultivated for agricultural and ornamental purposes. The rootstock and the leaves are poisonous in the raw state and must be cooked before eating. Wild taro is closely related to the ornamental elephant-ear, *Caladium*.

127
Growth aspect in flooded swamp

128
Large arrow-shaped leaf with undulate margins

Orontium aquaticum L.

Golden Club

Figures 129-131

129
Growth aspect of flowering plants with ovate leaves

130
Bright yellow flowering spadix

131
Green berry-like fruits on scape

Golden club is a perennial plant that grows from a rhizome deep in the substrate. The basal leaves are on petioles from 4-15 inches long. The blades are entire, elliptic to ovate and generally 8-9 inches long and 3 inches wide. The blades can vary from 4-15 inches in length and $1\frac{1}{2}$-$5\frac{1}{2}$ inches wide. The upper surface of the blade is blue green or dull green in color. The lower surface is usually silky smooth and silvery and the wide midvein is evident.

The inflorescence is a spadix that is borne on a scape from 5-20 inches long. The lower portion of the scape is lax and dull brownish-purple in color. The upper portion ascends while in flower and is usually white in color. This white portion of the scape contrasts greatly with the bright yellow flowers of the spadix. The spadix contains numerous flowers and is $1\frac{1}{4}$-$4\frac{1}{4}$ inches long. The 1-seeded berry-like fruits are green and $\frac{1}{3}$ inch wide.

Golden club flowers from late February to early April and mature fruits can be found in May. It grows in swamps and bogs and along streams and rivers throughout South Carolina but is most commonly found in the coastal plain.

Peltandra virginica (L.) Schott & Endlicher

Arrow Arum, Green Arum

Figures 132-134; Plates V-F, V-G

Arrow arum is a perennial, monoecious plant that grows from a thick rootstock. The leaves are clustered basally on a very short stem and the petioles reach 2 feet in length. The blades are 10-18 inches long, sagittate or hastate and have rounded or pointed basal lobes that are often widely flared. The blades have 3 main veins consisting of a central midrib and 2 veins that extend into the lobes. One or more small veins are adjacent to and parallel the entire leaf margin.

The flowers are arranged on a cylindric spadix with a green spathe enclosing the pistillate flowers below and enveloping the staminate flowers above. The fruiting spadix is a cluster of green berries enclosed by the spathe. Arrow arum flowers in May and June and fruits can be found throughout the summer and fall.

A closely related species, *Peltandra sagittifolia*, occurs only in the outer coastal plain. This rare plant is readily distinguished from arrow arum by its white spathe and red berries.

Arrow arum grows in swamps, marshes and bogs either on moist soil or in shallow water; it is found throughout the state.

132
Basally clustered leaves with long petioles

133
Leaf with prominent midrib and long flaring basal lobes

134
Slightly open green spathe revealing staminate portion of spadix

135
Growth habit of submersed plants

136
Plants with spirally arranged, short, linear leaves

137
Enlargement of small pink-purple flower

Mayacaceae

Mayaca fluviatilis Aublet

Bog Moss

Figures 135-137

Bog moss is a small, dark green, branched, aquatic or wetland plant. The stems are usually prostrate and root at the nodes; however, they can also be found completely submersed in water. The leaves are bright to dark green, spirally arranged on the stem, linear, and $^1/_8$ -$^1/_2$ inch long.

The flowers are produced singly on long, axillary pedicels. The flowers are showy, but quite small; the 3 pink to light purple petals are little more than $^1/_8$ inch long. The fruit is a several-seeded capsule that is only $^1/_5$ inch long.

Bog moss usually forms creeping mats on moist soil and strongly resembles club moss. When growing in marshy areas it is often inconspicuous as it is overtopped by other vegetation. When it grows completely submersed, the stems are usually longer and trail in the water. Plants growing on soil are sometimes referred to as *Mayaca aublettii*; however, it is thought that the wetland form and the aquatic form are probably the same species.

Bog moss flowers from late spring through the summer and is common in the coastal plain.

Commelinaceae

Murdannia keisak

(Hasskarl) Hand.-Mazz.

Asian Spiderwort,
Asian Dayflower

Figures 138-140; Plate V-E

This weak-stemmed and somewhat succulent annual is an introduction from eastern Asia. It was previously known as *Aneilema keisak* Hasskarl. The stem is readily decumbent and roots extensively at the nodes. Eventually, the ascending portion may gain a height of 18 inches. The leaves are alternate, narrowly lanceolate, and up to 3 inches long with parallel veins. The base of the leaf clasps the stem and becomes a sheath around the stem to ° inch long.

The flowers are axillary on short pedicels to 1/4 inch long; they are either solitary or there can be 2-4 flowers in a raceme. The 3 greenish sepals are smaller than the petals. There are 3 pink to lavender petals that are about 1/3 inch long. The capsule is also about 1/3 inch long and several-seeded. Flowers are produced in September and October.

Asian spiderwort grows in swamps, marshes, ditches, along stream banks and shallow margins of lakes and ponds, and is occasionally submersed; extensive colonies sometimes develop in the late summer. It is found throughout the state.

138
Growth aspect of flowering plants

139
Alternate, narrowly lanceolate leaves with clasping bases

140
Enlargement of three-petaled flower

Pontederiaceae

Pontederia cordata L.

Pickerelweed

Figures 141-142; Plate V-H

141
Growth aspect in shallow water

142
Showy many-flowered spike

Pickerelweed is an erect, perennial plant with basal leaves. It grows to 3 feet tall and arises from short thick rhizomes. The leaves are extremely variable in shape and two forms have been recognized based on leaf shape. The form with lanceolate blades and a narrow base has been recognized as *Pontederia cordata* var. *lancifolia.* The form with triangular-ovate or triangular-lanceolate leaves and a truncate to strongly cordate base is *P. cordata* var. *cordata.* The leaf blades are leathery, entire, rounded at the apex, parallel veined, and from 6-12 inches long. The petioles extend downward into a long sheath. The stem bears a bracteal leaf and a bladeless sheath from which the inflorescence arises.

The flowers are numerous and showy, arranged in a spike from 2-6 inches in length. The flowers and the axis of the spike are usually pubescent, often with glandular hairs. The flowers are irregular in shape and the perianth is united below to form a tube. The 6 perianth lobes are blue and the central upper lobe is marked with 2 yellow spots. There are 6 stamens, 3 are long and exserted from the perianth, and 3 are short. The fruit is about 1/4 inch long, reddish, glutinous and 1-seeded.

Pickerelweed flowers from April to October and grows in ditches and along margins of lakes, ponds and streams, often in water up to 2 1/2 feet deep. Pickerelweed is found throughout the state.

Amaryllidaceae

Hymenocallis spp.

Spider-lilies

Figures 143-146

Spider-lilies are perennial plants with basal leaves that arise from large bulbs. The leaves are linear, 1-3 feet long and about $^1/_2$-$1^1/_2$ inches wide. Several showy, white, sessile flowers are produced in an umbel at the end of a leafless stem. The six perianth parts are linear, 2-4 inches long and united below into a long, slender tube, that is also 2-4 inches long. Within the perianth is a showy white corona or cup-like structure that connects the bases of the anthers. The anthers extend well beyond the corona. The fruit is a few-seeded capsule.

The taxonomy and nomenclature of *Hymenocallis* in our area is in a state of uncertainty and is currently under study. One species of extreme interest is the rocky shoals spider-lily, *H. coronaria*, which is considered to be imperiled in South Carolina and is a candidate for federal listing. It only occurs in rocky shoal areas along the fall line in large piedmont rivers. Other species are found in swampy woods in the upstate and in swamps and tidal freshwater marshes in the coastal plain.

143
Hymenocallis coronaria *colony in rocky shoals*

144
Growth habit of ***H. coronaria***

145
Showy white flowers of ***H. coronaria*** *with cup-like corona*

146
Growth habit of ***H. crassifolia*** *in a low country marsh*

Orchidaceae

Habenaria repens Nuttall

Water-spider Orchid

Figures 147-149

147
Flowering plants growing in a floating mat of frog's bit

148
Linear-lanceolate leaves with sheathing bases

149
Details of flowers in raceme

This orchid is a perennial plant that grows from tuberous roots. It is succulent, stoloniferous, grows to 2 feet tall, and is usually light to medium green in color. The leaf bases sheathe and entirely cover the stem; the blades are oblong or linear-lanceolate, strongly 3-ribbed, up to 10 inches long and 1 inch wide.

The inflorescence is a densely flowered raceme from 4-8 inches in length. The flowers are bracteate, with green to pale yellow slender perianth parts and a long spur. The fruit is a many-seeded capsule.

The water-spider orchid flowers throughout the growing season and often grows in dense stands in shallow water of pond and lake margins, marshes or in floating mats of vegetation. It can be found in the piedmont but is primarily a species of the coastal plain.

Saururaceae

Saururus cernuus L.

Lizard's Tail

Figures 150-152; Plates II-I, IV-H

Lizard's tail is an aquatic and wetland perennial plant that spreads from rhizomes. The stems grow to 3 feet tall and are somewhat pubescent. The leaves are alternate, dark green above and lighter green below, heart-shaped, and up to 5 inches long.

The flowers are arranged in curving racemes that arise opposite the upper leaves; the tip of the raceme is usually pointed downward. The racemes are 4-8 inches long, about 1/2 inch wide, and are borne on peduncles up to 3 inches long. There are no sepals or petals; the showy aspect of the flowers is from the 3-6 white stamens and pistils. The fruit is a few-seeded capsule.

Lizard's tail flowers in late spring and summer. Plants usually occur in sizable populations in low woods, water margins, swamps and marshes. This species occurs throughout the state except at higher elevations.

150
Growth aspect in shallow water habitat

151
Curved tail-like flowering racemes

152
Dark green heart-shaped leaves

Urticaceae

Boehmeria cylindrica (L.) Swartz

False Nettle

Figures 153-155

False nettle is a highly branched, monoecious, perennial plant reaching 3-4 feet in height. Although herbaceous, it often has a shrub-like appearance. The leaves are opposite, ovate, long-petiolate, and up to 8 inches long. The blade has 3 main palmate veins from the base and a serrate margin. Two types of leaves are produced by false nettle. Those plants growing in the shade tend to have larger, thinner leaves, whereas those growing in full sun have somewhat smaller, thicker leaves.

Small male and female flowers occur in small clusters on short branches in the axils of the leaves. The flowers are often obscured by the leafy branches. The fruit is an achene.

False nettle can be found in flower from July to August and the achenes are produced in the early fall. It grows in marshes and swamps, often occurring as a main component of the floating log and stump vegetation. It is found throughout the state.

153
Shrub-like growth on stump in flooded swamp

154
Ovate leaves with serrate margins

155
Small flower clusters on lower portion of leafy branches

Polygonaceae

Polygonum densiflorum Meissner

Dense-Flower Smartweed

Figures 156-158

Dense-flower smartweed is a robust glabrous perennial from a stout rootstock. The stems are decumbent and root at the nodes, then ascend and branch, reaching lengths of up to 8 feet. The stems are terete, up to $^3/_4$ inch in diameter at the nodes and dark green to reddish brown in color. The leaves are alternate, smooth and entire. The blade is up to 10 inches long and to $3^1/_2$ inches wide, elliptic to widely lanceolate, pointed at both ends, and attached to a petiole varying from $^1/_4$-$1^1/_2$ inches long. The ocrea is not ciliate at the apex and is up to $1^1/_4$ inches long.

Flowers are borne in numerous erect, axillary racemes that are 2-4 inches long. Petals are absent, the sepals are pink or occasionally white. The fruit is an achene that is dark brown, lenticular, and about $^1/_8$ inch long.

Dense-flower smartweed flowers from July to October and grows in lakes and swamps in water up to 4 feet or more deep. It often forms dense stands and is frequently associated with *Hydrocotyle ranunculoides*. It is usually found in the coastal plain.

156
Dense flowering stand along river's edge

157
Numerous flowering racemes

158
Detail of ocrea and leaf petiole

Polygonum lapathifolium L.
Willow-weed, Nodding Smartweed
Figures 159-161; Plates III-D, III-G

159
Dense stand at river's edge

160
Drooping racemes

161
Inflated stems rooting at the nodes

This species of smartweed is an introduction from Europe and is similar to *P. densiflorum* except it is an annual plant and the flowers are usually white. On robust plants, the lower stems become inflated to 1 inch or more in diameter; these lower internodes become partially hollow and spongy inside.

Flowers are produced in numerous, narrow, drooping racemes. When in flower, the drooping white racemes of willow-weed can resemble those of lizard's tail from a distance. Individual flowers are usually white and are less than $^{1}/_{8}$ inch long. The fruit is a dark brown, lenticular achene that is usually $^{1}/_{10}$ inch long or less. Flowers and fruits are produced from mid-summer through the fall.

Willow-weed grows along marshy margins of streams, rivers, lakes, and other wet areas, frequently occurring in large, dense stands. This is presumably a recent introduction to this state; it was not known to occur in South Carolina in the 1960's but is now found throughout the state.

There are several other species of *Polygonum* that occur in moist, wet or aquatic habitats; most of these occur throughout the state. They are distinguished by the size and type of margin of the ocrea, the color of flowers, the presence or absence of glands on the flower parts, and the presence or absence of downwardly pointing barbs on the stems, among other features.

Amaranthaceae

Alternanthera philoxeroides (Martius) Grisebach

Alligatorweed

Figures 162-165; Plates I-H, II-F

This aggressive introduction from South America is a mat-forming perennial that spreads rapidly in aquatic or wet situations by seed production and by fragmentation. The stems reach lengths of 3 feet or more, root extensively at the nodes and, except for the ascending tips, are usually decumbent, trailing or floating. The stems are light to dark green or often pink to dull reddish-purple, hollow and usually $^{1}/_{4}$ inch or less wide. The internodes often become inflated and reach $^{1}/_{2}$ inch in diameter. The leaves and stems are fleshy and somewhat succulent.

The leaves are opposite, entire, oblanceolate to elliptic and to 4 inches long. The midrib is prominent, especially below, and the blades are sessile or tapering at the base. The leaf bases form a short sheath at the stem; therefore, the nodes are prominent.

Small white flowers are borne in short, headlike spikes on peduncles $^{1}/_{2}$-3 inches long arising from the upper leaf axils. These spikes are $^{1}/_{4}$-$^{1}/_{2}$ inch wide. Only 1 spike is produced per node; when more than 1 spike is present on a flowering stem, there is at least 1 node between them. Petals are lacking; there are 5 separate sepals that are white or colorless and about $^{1}/_{4}$ inch long. The fruit is 1-seeded.

Alligatorweed flowers from April through October and grows anywhere moist or wet conditions prevail – lakes, streams, ditches, and floating mats of vegetation. It can form large, dense mats that may impair access and use of a waterbody. Alligatorweed is found in the piedmont and coastal plain.

162
Extensive growth in river

163
Aggressive mat-forming growth restricting access

164
Growth habit of young, flowering plants

165
Clover-like head of white flowers arising from leaf axil

Callitrichaceae

Callitriche heterophylla Pursh emend. Darby

Water-starwort

Figures 166-168

166
Growth aspect of plants with floating leaves in shallow water

167
Heart-shaped fruits in axils of emersed, opposite leaves

168
Tangled submersed growth with linear leaves

Water-starwort is an annual aquatic plant that usually grows as a submersed, tangled mat with few to many floating and emersed leaves. The submersed leaves are 1-nerved, entire, linear and $^1/_2$ -1 inch long. The floating and emersed leaves are opposite or crowded in rosettes and spatulate to obovate in shape. These leaves are 3- to 5-nerved, entire and usually $^1/_4$-$^1/_2$ inch long. Occasionally, only submersed leaves are present. Diminutive flowers and heart-shaped fruits are formed in the axils of the leaves.

Water-starwort grows in quiet water of streams, swamps, lakes, flood plains and ditches. It usually occurs as a submersed plant with floating leaves, but can be found entirely submersed or emersed. Water-starwort is abundant in late winter and early spring but can be found throughout South Carolina almost year-round.

Onagraceae

Ludwigia spp.

Primroses

Figures 169-172; Plates I-I, IV-A, IV-E

The primroses are a large group of wetland plants that share several common features. The flowers usually have 4-6 bright yellow petals and a corresponding number of sepals. The fruit is a many-seeded capsule that is inferior in position, i.e., formed below the point of attachment of the sepals and petals. The capsule is topped by the persistent sepals.

However, primroses are quite variable. Combinations of the following characteristics are used to identify the species. Most species of primrose are perennials but a few are annuals. The stems can be erect or prostrate and the lower stems of several species become white and spongy-thickened. The stems of a few species are wing-angled due to decurrent leaf bases. The leaves can be opposite or alternate, sessile or petioled, and glabrous or pubescent. Most species have 4-6 petals; however, the petals are absent in several species. The capsular fruits can be winged, 4-angled, ribbed or smooth. These capsules can be sessile or pedicellate and glabrous or pubescent. The capsules can vary in shape from subglobose to cubical, cylindric, obconic or obpyramidal.

Primroses are usually found in flower from May through October. They grow in marshes, ditches, bogs, savannas, pine flatwoods and shorelines of lakes and rivers. At least 23 species of *Ludwigia* are known to occur in South Carolina. Several of these are rare or infrequently encountered. Primroses can be found throughout South Carolina but the majority of the species are restricted to the coastal plain.

169
Glabrous stem, leaves and 4-petaled flower of ***Ludwigia alternifolia***

170
4-petaled flower of ***Ludwigia brevipes***

171
5-petaled flower and pubescent stem of ***Ludwigia leptocarpa***

172
Elongate pubescent capsule of ***Ludwigia leptocarpa***

173
Extensive growth in lake habitat

174
Early rosette stage growing out over water surface

175
Detail of ovate shaped leaves of rosette stage

Ludwigia hexapetala (Hook & Arn.) Zardini, Gu & Raven

Water Primrose

Figures 173-177

Formerly known as *Ludwigia uruguayensis*, water primrose is a perennial aquatic plant that exhibits two different growth forms, referred to here as the vegetative or rosette stage and the flowering stage.

The vegetative stage is characterized by glabrous to sparsely pubescent stems that grow horizontally over the soil or water and produce roots at the nodes. These roots typically are feathery in appearance but wiry in texture; white, spongy aerenchymous roots are often produced at the nodes. The leaves are produced alternately along this horizontal stem or are alternate on much-shortened vertical branches formed at the nodes. The vertical branches have very short internodes such that the leaves appear to be in rosettes. The leaves in the vegetative stage are essentially glabrous and vary in shape from suborbicular to obovate to spatulate; the apex is rounded. The veins are usually very distinct and sometimes are tinged reddish-purple. The margins are entire and the petioles vary from $^{1}/_{16}$-2 inches long.

The flowering stage is characterized by erect stems, which result from the elongation and growth of the 'rosette' branches. The flowering stems are usually villous and branched, growing to 3 feet tall. The leaves are variable, from lanceolate to elliptic in shape; both ends are usually pointed. They are sessile or on short petioles and are sparsely pubescent and ciliate along the margins.

The flowers are very showy and are produced from late April to September. They arise singly on pedicels 1-1 1/2 inches long from the axils of the leaves. The pedicels and outer surface of the sepals are covered with white villous hairs. The ovary is inferior, situated below the sepals and petals. There are 5 or 6 separate narrow sepals up to 3/4 inch long. The petals are bright lemon-yellow, strongly veined, separate, widely obovate, 5 or 6 in number, and 1 inch long. The fruit is a many-seeded cylindrical capsule to 1 inch long, capped by the persistent, spreading sepals. By late summer the erect flowering stems become tough and red in color.

By mid to late fall, the flowering stems become partially decumbent and root at the nodes. The vegetative stage over-winters with very small leaves present. In late winter to early spring, growth begins anew.

Water primrose grows along the margins of rivers, lakes and ponds where it often forms extensive floating mats that impair water use activities. It is found throughout South Carolina.

176
Upward elongation of maturing plants

177
Alternate lanceolate leaves of young flowering stage

178
Early growth stage along shoreline

179
Red-purple form in early growth stage

180
Detail of opposite, ovate leaves

Ludwigia palustris (L.) Elliott
Creeping Primrose
Figures 178-180

Creeping primrose is a prostrate, mat-forming perennial that roots at the nodes and is glabrous. The stems are often red in color. The opposite leaves vary in color from entirely green to green above and red below, or they can be red-purple tinged. The leaf blades are elliptic to ovate and 1/2-1 inch long; they narrow at the base to petioles that are often as long as the blades.

Small sessile flowers without petals occur in the axils of the leaves from late spring through the fall. The small 4-angled capsular fruits are less than 1/8 inch long and are topped by the 4 persistent sepals. These capsules have longitudinal green stripes.

Creeping primrose grows in moist areas, along shorelines and in shallow water in a variety of habitats. It can also be found submersed in shallow water of lakes, ponds and ditches. Creeping primrose occurs throughout the state.

Ludwigia spathulata and *L. repens* are two species of primrose that are quite similar in appearance to *Ludwigia palustris* but are considered rare or uncommon in South Carolina. In contrast to the common *L. palustris, L. spathulata* is pubescent. The flowers of *L. repens* have 4 yellow petals and the capsule lacks the longitudinal green stripes of *L. palustris.*

Apiaceae

Cicuta maculata L.

Water Hemlock

Figures 181-183; Plate I-E

Water hemlock is a robust, glabrous perennial that grows from tuberous-like roots. The stems are hollow and ribbed, to 6 or 8 feet tall and they are often mottled with purple below. The leaves are mostly 1-3 times pinnately compound with long, clasping petioles. The lower leaves are up to $1^1/_2$ feet long, while the upper leaves become much smaller. The leaflets are usually lanceolate with serrate or dentate margins.

The flowers are in lateral and terminal compound umbels; each umbel is 2-6 inches across and consists of numerous, very small white flowers. The fruits are small and nutlet-like.

These plants are **deadly poisonous** to humans and animals. Water hemlock flowers in the summer; it grows in marshes and along margins of swamps, lakes, rivers, and ditches throughout the state. It is often found growing on well-developed floating log or stump communities or on small tree islands.

Another species of water hemlock, *Cicuta mexicana*, also occurs in South Carolina. Plants of *C. mexicana* are stouter and larger than *C. maculata*. The large stems are strongly ribbed with alternating green and purple stripes and the leaflets are ovate.

181
Growth habit in swamp habitat

182
Compound leaves with serrate margins and clasping petiole

183
Many-flowered umbel

Hydrocotyle ranunculoides L.f.
Water Pennywort
Figures 184-186

184
Floating mats in swamp habitat

185
Dense growth of dark green, orbicular leaves

186
Detail of leaf with irregularly crenate margin and central red spot

Water pennywort is a succulent, glabrous perennial that usually forms dense mats along shores or grows in floating mats. The stems are stout, up to 1/4 inch in diameter, decumbent and they root extensively at the nodes, even when floating. The alternate leaves have petioles that vary in length up to 12 inches long. The blades are irregularly crenate, shallowly 3-5 lobed, and cleft at the base to the petiole. The blades are up to 3 inches wide, usually slightly wider than long. The upper surface is bright green, often with a central reddish-purple spot at the point of petiole attachment; the lower surface is light green.

Small white flowers with 5 petals are produced in simple few-flowered umbels. The peduncles that support these umbels are less than 1/2 the length of the petiole of the subtending leaf.

This species often forms large floating mats in swamps and lakes. These floating mats can support other plants such as *Bidens* spp., *Polygonum densiflorum*, *Alternanthera philoxeroides*, and *Myriophyllum aquaticum*. Submersed aquatic plants are often found tangled in the roots. Extensive populations can occur during the fall, winter and spring but, after flowering in April and May, they generally die back during summer. Water pennywort is restricted to the coastal plain.

Hydrocotyle umbellata L.
Marsh Pennywort
Figures 187-189; Plates III-H, IV-C

Marsh pennywort is a glabrous perennial with slender, decumbent stems that root at the nodes. The leaf blades are peltate, orbicular and 1-2 inches wide; the margins are crenate. Petioles are usually no more than 6 inches long.

Small, white flowers are produced in simple, many-flowered umbels. The umbels are supported on peduncles that are usually as long as or longer than the leaves. Flowering occurs from late spring to early fall.

Marsh pennywort is usually found in marshes, in shallow water along margins of lakes and ponds, or on moist soil. It can be found throughout the piedmont and coastal plain.

Whorled pennywort, *Hydrocotyle verticillata*, has leaves very similar to those of marsh pennywort. The flowering stems differ in that the flowers are produced in several whorls or verticils rather than in a simple umbel. The flowers and fruits may be either sessile or pedicelled in these whorls. Whorled pennywort is typically found in forested wetlands in the coastal plain and piedmont.

187
Orbicular leaves and umbels of flowers

188
Growth habit of flowering plants

189
Detail of peltate, orbicular leaf with crenate margin

Haloragaceae

Myriophyllum aquaticum (Vell.) Verdc.

Parrot-feather

Figures 190-192

190
Submersed and emergent growth habit

191
Flowers formed in axils of leafy bracts

192
Detail of feathery whorled bracts

Previously known as *Myriophyllum brasiliense*, parrot-feather is a perennial plant introduced from South America. The stems are often tinged with orange or red. It grows rooted and submersed in up to 3 feet of water and forms numerous adventitious roots; the main branches are emersed. The emersed leaves are in whorls of 3-6, usually pale green to gray-green, $^{1}/_{2}$-2 inches long and to $^{1}/_{2}$ inch wide. The leaves are divided into very narrow segments, appearing feathery. The submersed leaves are larger than the emersed leaves but the capillary segments are narrower and not as firm in texture.

Only pistillate plants occur in the United States. The flowers are inconspicuous and are formed in the axils of the emersed leaves in the spring. However, parrot-feather is rarely found in flower.

Parrot-feather is most commonly found in the late fall, winter and early spring and often forms large, extensive mats on the surface of the water. Alligatorweed and water pennywort are often associated with this species. Parrot-feather grows in lakes, ponds, swamps, and along stream margins throughout the state.

Hydrophyllaceae

Hydrolea quadrivalvis Walter

Hydrolea, Water-pod

Figures 193-194

Hydrolea is a perennial plant that grows in water up to 6 feet deep. The unbranched, pubescent stems are decumbent and root at the nodes. Submersed leaves, when present, are thin and flaccid. The emersed leaves are alternate, entire, and lanceolate to elliptic in outline. They are usually 2-5 inches in length and have a short petiole. The leaves are usually subtended by a sharp thorn that is up to 1/2 inch long.

The flowers are single or in clusters in the upper leaf axils. They are bright blue and about 1/3 inch wide. The fruit is a subglobose capsule filled with many very small seeds.

Hydrolea flowers throughout the summer and grows in swamps, marshes, ponds, and lakes. Early in the year it can be found entirely submersed, but this is an uncommon occurrence. Hydrolea occurs in the lower piedmont and throughout the coastal plain of South Carolina.

193
Mature plant with alternate elliptic leaves, flower and fruit

194
Details of flowers

Scrophulariaceae

Bacopa caroliniana (Walter) Robinson

Lemon Bacopa, Blue-hyssop

Figures 195-197

195
Growth aspect at edge of water

196
Dense growth habit in shallow water

197
Details of flower and opposite ovate leaves

Lemon bacopa is a succulent perennial that grows from rhizomes. The common name is derived from the aromatic, lemony scent given off when the leaves are crushed. The plants are usually creeping and mat-forming on moist soil, or they can be found as emergent plants in shallow water. The stems are pubescent and often stout. The leaves are opposite, glandular-punctate, elliptic to ovate, and 1/4-1 inch long with the base clasping the stem. The leaves are sessile, palmately veined, often dark green when submersed and bright green when emersed.

The flowers are blue, 1/3-1/2 inch long and axillary on short pedicels; pedicels and flowers are no longer than the length of the subtending leaf. The pedicels become elongate when in fruit, reaching two times the length of the subtending leaf. The fruit is a capsule.

Bacopa flowers throughout the summer and fall. It grows in shallow water or along margins of ponds and lakes and in bogs and ditches throughout the state.

Micranthemum umbrosum (J. F. Gmelin) Blake

Shade Mudflower

Figures 198-200

Shade mudflower is a small, bright green, glabrous plant. The creeping stems root at the nodes and form mats on moist soil. The leaves are opposite, sessile, entire and palmately veined. The leaves are orbicular and $^{1}/_{6}$-$^{1}/_{3}$ inch long and wide. Submersed leaves are usually more flaccid and larger than those that are emersed.

Solitary flowers, each comprised of 4 sepals and 4 white petals, are formed in the axils of the leaves. These flowers are about $^{1}/_{20}$ inch long and are easily overlooked because of their minute size.

Micranthemum grows on moist soil in bottomlands, along margins of lakes and ponds, or it often can be found entirely submersed in rivers and streams growing in bright green mats attached to logs or stumps. It is most commonly found in the coastal plain, although it does occur in the piedmont.

198
Growth aspect in shallow water

199
Emergent growth on moist soil

200
Floating mat attached to other aquatic plants

Acanthaceae

Justicia americana (L.) Vahl

Water Willow

Figures 201-203

201
Small colony of flowering plants

202
Opposite narrowly elliptic leaves and flower spike

203
Details of flower with conspicuous lower lip

This rhizomatous perennial usually forms colonies or dense stands. The stems are glabrous, prominently jointed and up to 30 inches tall. The leaves are opposite, narrowly elliptic, lanceolate, or linear, entire and up to 6 inches long. The leaf blade narrows to the base or to a short petiole.

The flowers are opposite on crowded spikes. The peduncles arise from the upper leaves and are often longer than the subtending leaves. There are 5 green narrow calyx lobes. The petals are united to form a $^3/_4$ inch long flower with a conspicuous lower lip. The petals are pale violet with purple markings, especially on the lower lip. The fruit is a $^1/_2$ inch long 4-seeded capsule that splits lengthwise to expose the seeds when mature.

Water willow flowers from May to October. It grows along rocky or sandy streams and in shallow water of lake margins in the piedmont and coastal plain.

Asteraceae

Bidens spp.

Beggar-ticks, Stick-tights,
Bur-marigolds

Figures 204-206

These are showy plants that grow to 3 feet or more in height. They have opposite leaves and yellow to yellow-orange heads of daisy-like flowers. The flower heads are subtended by 2 series or whorls of bracts; the outer bracts are green and leafy and the inner bracts are yellowish or brownish. Many achenes or nutlets are produced by each flower head. The nutlets are flat, about $^{1}/_{4}$ inch long and topped by 2 downwardly barbed awns. The awns readily stick to clothing or the fur of animals, giving meaning to the common names.

There are at least 8 wetland species of *Bidens* in South Carolina. Most species are annual plants; however, *B. laevis* is a rhizomatous perennial. The leaves vary greatly from simple to dissected to compound. The flowers are often very showy, reaching up to 3 inches wide in some species.

The wetland species of *Bidens* can usually be found in flower and fruit from September through November providing a late-season show of color at a time when other plants are beginning to die back. Some species form mats that float on the surface of the water or they grow in mats formed by other plants. They grow in marshes, swamps and wet ditches throughout South Carolina.

204
Flowering mats growing in a swamp habitat

205
Stand of ***Bidens*** *along shoreline*

206
Details of daisy-like flowers of ***Bidens***

IV. Grasses, Sedges and Rushes

Introduction to Grasses, Sedges, and Rushes

Graminoids, as grasses, sedges and rushes are often called, are ubiquitous in aquatic and wetland environments. While these plants may superficially appear similar, they can be distinguished by a few basic characteristics.

Poaceae - grasses usually have hollow, round stems with 2-ranked leaves. The leaf blades are linear and the sheaths are usually open with overlapping margins. The flowers consist of bract-like scales enclosing the pistil and stamens. The fruit is a grain, also known as a caryopsis.

Cyperaceae - sedges have solid, triangular stems with 3-ranked leaves. The leaves consist of a linear blade and a fused sheath although some have leaves consisting of a sheath only. Individual flowers are composed of a pistil, stamens, and usually, bristles or small scales. Each flower is subtended by 1 or rarely 2 bracts. The fruits are small nutlet-like achenes.

Juncaceae - rushes have round stems and leaves The leaves, consisting of either a sheath or a sheath and a blade, are usually basal or low on the stem. The leaves may or may not be septate and the blades, when present, are flattened or terete. The flowers are 3-parted and scale-like. The fruit is a capsule with numerous very small seeds.

The grass family is the second largest family of flowering plants in the world and, of the approximately 300 grass species in South Carolina, almost half are found in aquatic or wetland habitats. Only the more commonly encountered aquatic grass species are described in this chapter. General descriptions of the sedge and rush families are also presented with individual treatments of some representative species.

Poaceae

Hydrochloa caroliniensis Beauvois

Carolina Water Grass

Figures 207-209

This weak-stemmed monoecious perennial is freely branched and roots at the nodes; the lower portion is usually submersed and the upper branches are tangled and floating. The stems grow to 2 or 3 feet long; the leaves are up to 2 inches long, narrow and glabrous.

The flowers are white, about 1/6 inch long and somewhat inconspicuous. The staminate flowers are produced at the ends of the branches with the pistillate flowers below them. Flowers are produced sporadically in the summer and fall.

Carolina water grass grows in water to 3 feet deep in lakes, ponds and swamps. It is also found in slow-flowing water and on moist soil along margins of lakes and ponds. This aquatic grass is restricted to the coastal plain.

207
Emergent growth in shallow backwater area

208
Narrow floating leaves

209
Small white flowers produced at branch end

Panicum hemitomon Schultz
Maidencane
Figures 210-211

210
Dense shoreline stand

This rhizomatous grass is a common shoreline emergent that frequently grows in dense stands. The stems root and branch from the lower nodes and grow to about $2^1/_2$ feet tall. The leaf blades are scabrous along the margins, 6-10 inches long, to $^1/_2$ inch wide near the base, and long tapering to the end.

The inflorescence is a narrow and rather 1-sided panicle to 8 inches long. The flowering panicle branches are strongly ascending, resulting in a width usually no greater than $^1/_4$ inch. The spikelets are perfect, lanceolate and $^1/_{10}$ inch long.

Maidencane can be found in flower from May through August and grows along margins and in shallow water areas of lakes and ponds, as well as in marshy and swampy areas throughout the coastal plain.

Most of the panic grasses, as members of the genus *Panicum* are called, are wetland or aquatic plants. A non-native species, torpedo grass, *Panicum repens*, has been found growing in South Carolina. It has been known to cause problems in Florida due to its prolific growth in aquatic and wetland systems.

211
Growth habit; a) strongly ascending flowering branches; b) detail of spikelets

Paspalum distichum L.

Knot-grass

Figures 212-213

Also known as Panama paspalum, knot-grass is a decumbent, perennial grass that spreads by stolons and rhizomes. The stems root at the nodes; the flowering stems ascend to 1 foot tall. The leaf sheaths are loose and usually purplish. Leaf blades reach $3^1/_2$-4 inches in length and are narrow and often folded or inrolled.

The inflorescence terminates the flowering stem and consists of a pair of ascending spike-like racemes 1-2 inches long. The spikelets overlap and are in 2 rows on 1 side of the rachis. They are ellipsoid, 1-flowered and $^1/_{10}$-$^1/_8$ inch long.

Knot-grass flowers from mid-summer through the fall. Knot-grass grows throughout the state on various moist or wet substrates along open lake margins, in marshes, and on floating mats of vegetation.

Bahia grass, *Paspalum notatum* var. *saurae*, is similar in appearance to knot-grass. Bahia grass also has paired racemes terminating the flowering stem; however, the racemes of bahia grass are $2^1/_2$-6 inches long and the leaves can reach 10 inches in length. Bahia grass is a weedy introduction, commonly found growing along roadsides, in fields and pastures, and in other disturbed areas.

212
Low growth habit along lake margin

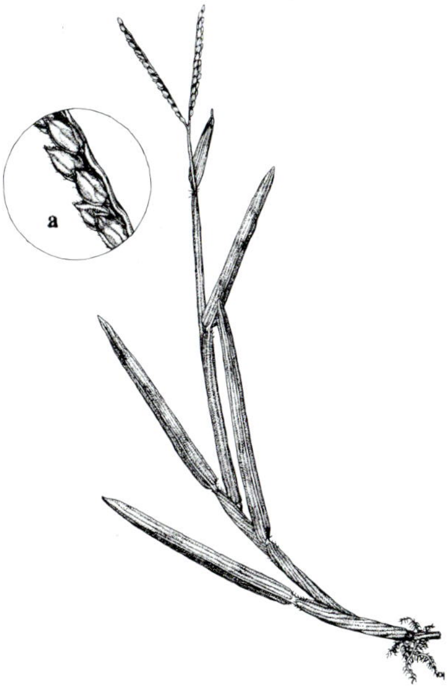

213
Folded leaf blades and paired spike-like racemes; a) spikelets on 1 side of the rachis

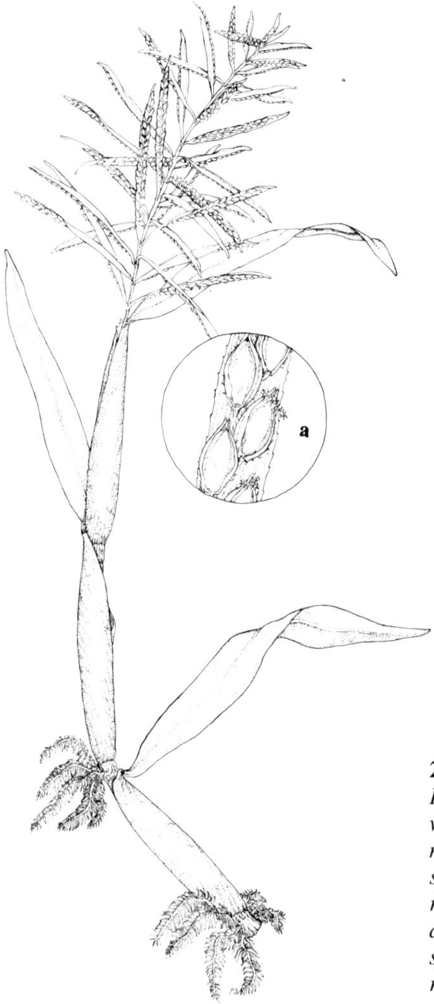

214
Inflated stems with rooting nodes and many spreading racemes;
a) detail of spikelets and rachis

Paspalum repens Berg.

Walter Paspalum

Figure 214

Formerly known as *Paspalum fluitans*, this is an annual grass with elongate spongy stems; it roots at the nodes on wet soil or is rooted in water to 6 feet deep or more. If in water, the stems produce roots at the nodes and the inflorescences and upper stems are emergent. The nodes appear as a dark band at the bases of the leaf sheaths. The leaf sheaths are loose; the blades are 6-8 inches long, $^1/_2$-$^2/_3$ inch wide and scabrous on both surfaces.

The inflorescence is up to 6 inches long, composed of 20-50 racemose racemes, or racemes appearing as branches on a central stalk. The racemes are spreading to ascending and up to 2 inches long. These racemes have many small, 1-flowered spikelets in 2 rows on 1 side of the rachis. The rachis is broad or winged and is as wide or wider than the 2 rows of spikelets.

Water paspalum flowers in September and October. It grows in low, wet areas or in relatively deep water in swamps in the coastal plain.

Several other species of *Paspalum* occur in South Carolina, many of which are wetland species.

Phragmites australis
(Cav.) Trin. ex Steud.
Common Reed, Phragmites
Figures 215-217

Previously known as *Phragmites communis*, common reed is a perennial grass that arises from thick, white, hollow rhizomes deep in the substrate. Once established, large colonies of this grass can occur, spreading readily by rhizomes. The slender stems reach 9-10 feet or more in height and are leafy up to the inflorescence. The leaves are distichous; the blades are blue-green to green, flat, long-attenuate, to $^1/_2$ inches wide and $1^1/_2$ feet long. The blades are glabrous with scabrous margins. Silky white hairs, $^1/_4$ inch or more in length, are present at the junction of the leaf sheath and blade.

The leafy stem is topped by a large tawny, brown, purplish or silvery panicle of flowers over 1 foot long. The flowers or spikelets become silvery-silky with age due to the long hairs associated with the spikelets.

Common reed flowers from mid-summer through the fall. In South Carolina, it is generally restricted to the outer coastal plain where it occurs in fresh, brackish and salt marshes and along streams, rivers and estuaries.

Phragmites is similar to the giant reed, *Arundo donax*. *Arundo* is distinguished by having a larger inflorescence that is tawny or light colored, and by a large clasping flange of tissue at the junction of the blade and sheath.

215
Extensive shoreline colony

216
Slender stems with flat, long, green to blue-green leaves

217
Large panicle of flowers

Sacciolepis striata (L.) Nash
Baggy-knees Grass

Figures 218-220

This perennial grass is stoloniferous with stems that become decumbent and root at the lower nodes. The glabrous, branched, vertical stems eventually attain heights of up to 3 feet. The leaves are glabrous, narrowly lanceolate, to 6 inches long and $^{1}/_{2}$-$^{2}/_{3}$ inch wide at the base. The blades taper to a long apex and the bases are rather cordate. The leaf sheaths are usually pubescent.

The inflorescence is a narrow to cylindric panicle, 3-7 inches long. The spikelets are glabrous and green, bulging on one side, 1-flowered, strongly nerved and about $^{1}/_{8}$ inch long.

Baggy-knees grass flowers in the summer and fall and grows along wet, open margins of lakes and ponds and in marshes. It often grows in water to 2 feet deep and it can form dense colonies. Baggy-knees grass is common throughout the state.

218
Growth aspect along open lake margin

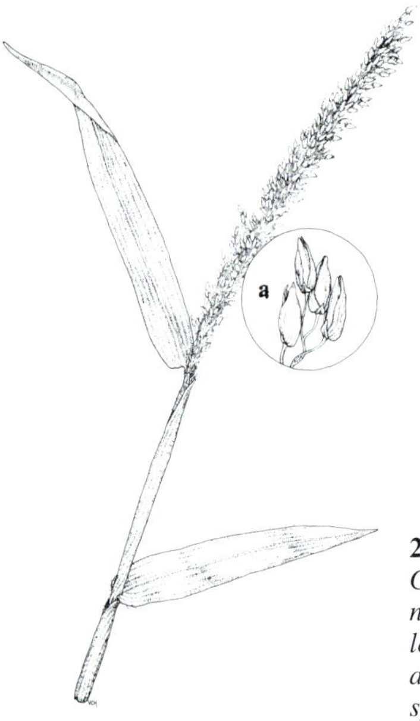

219
Growth habit, note cordate leaf bases; a) enlarged spikelets

220
Cylindrical panicle with smooth, green spikelets

Zizania aquatica L.
Wild Rice
Figures 221-223

Wild rice is a monoecious, rhizomatous, perennial grass. The stems, which are up to 10 feet tall, are often decumbent below and root at the nodes. The stems, leaves and sheaths are glabrous. The leaf blades are flat with scaberulous margins and can reach 3 feet or more in length and 2 inches in width.

The inflorescence is a terminal panicle that is quite striking in appearance and size. Panicles can reach up to 2 feet in height and the lower portion can be up to 1 foot wide. The upper portion consists of strongly ascending branches that bear the pistillate spikelets. The pistillate spikelets are appressed to the rachis, ascending, pubescent, awned, and 1/2 inch or more long. The lower portion consists of spreading branches that bear the staminate spikelets. These spikelets, which are usually no more than 2/3 inch long, bear bright yellow stamens and readily drop from the plant when they are finished flowering. Therefore, when the plant is in fruit, the lower branches of the inflorescence appear naked.

Wild rice can be found in flower from spring until the fall, but the peak flowering season is in the spring. It grows in freshwater and brackish marshes and along streams and rivers in the coastal plain.

221
Small colony with light green inflorescences

222
Inflorescence with light green pistillate branches above and tawny staminate branches below

223
Staminate flowers

Zizaniopsis miliacea (Michaux) Doell & Ascherson

Giant Cutgrass, Southern Wild Rice

Figures 224-226

224
Large, dense stand

225
Enlarged view of "cutting" leaf margins

226
Staminate flowers

This is a large, robust monoecious perennial grass that grows up to 10 feet tall and usually forms dense stands. The rhizomes are large and scaly. The stems are coarse, leafy and unbranched; the lower stems often reach 1 inch in diameter and root at the nodes. The stems, leaves and sheaths are glabrous; the lower surface of the blade is often scaberulous and the margins of the blades are upwardly scabrous, hence the common name, cutgrass. Leaf blades have a prominent midrib and are usually $1^1/_2$ inches wide or occasionally as wide as $2^1/_2$ inches.

The inflorescence is a large, open panicle, from 1-2 feet high. The branches are ascending, each branch with pistillate spikelets on the outer portion and staminate spikelets on the lower or inner portion of the branch. The pistillate spikelets are 1-flowered, about ˘ inch long, rounded and purplish, with an awn almost as long as the spikelet.

Giant cutgrass can be found in flower from April to July. It grows in marshes and along stream and lake margins in shallow water to 3 feet deep. It occurs in the coastal plain and the lower piedmont.

Cyperaceae

Sedge Family
Figures 227-229

The sedges are a group of grass-like herbaceous plants characterized by usually having solid, triangular stems and alternate, 3-ranked leaves. Each leaf consists of either a blade and a closed sheath or simply a bladeless, closed sheath, as in the spikerushes.

The flowers are very small and lack a perianth, or in some species, the perianth is present as bristles or scales. The flowers are subtended by scales and they occur in clusters called spikelets. It is mainly the characters of the flower and spikelet upon which the species of sedges are distinguished by plant taxonomists. However, many of the sedges do have a distinctive appearance, thus allowing for on-sight identification.

There are 16 genera and over 250 species of sedges in South Carolina, the majority of which are considered to be aquatic and wetland plants. Some of the more commonly encountered genera in aquatic habitats are *Carex, Cyperus* (nutgrasses), *Eleocharis* (spikerushes), *Dulichium* (three-way sedge), *Rhynchospora* (beak-rushes), and *Scirpus* (bulrushes).

227
Inflorescence of ***Cyperus erythrorhizos***

228
Three-ranked leaf arrangement of ***Dulichium arundinaceum***

229
Inflorescence of ***Rhynchospora corniculata***

230
*Long narrow leaves and yellow-green spikes of **Carex comosa***

231
Carex decomposita *(Cypress-knee sedge) growing on a cypress stump*

232
Carex lupulina *inflorescence*

233
Carex glaucescens *inflorescence*

Carex spp.
Sedge
Figures 230-233

At least 60 species of *Carex* in South Carolina are considered to be wetland species. They are perennial, usually grow from rhizomes, and the stems are terete. In most species, the stems are usually no more than 2 or 3 feet tall. The leaves are 3-ranked and consist of both sheaths and blades.

Details of the spikes, scales, flowers, and mature achenes are needed to identify the species of *Carex*. *Carex* is monoecious; separate male and female flowers are formed on each plant. In one group of *Carex,* all spikes are similar with the upper flowers of the spike all one sex and the lower flowers are of the opposite sex. In the other group, staminate flowers are found on one spike, usually the uppermost spike, and pistillate flowers on the other spikes.

Individual flowers are subtended by scales. Staminate flowers consist only of 3 stamens each. Pistillate flowers are made up of a pistil and a sac-like structure, the perigynium, which encloses the pistil. The fruit is an achene which can be either lenticular or 3-angled.

Most species of *Carex* flower in the spring or early summer. The wetland species can be found in swamps, marshes, shorelines, floating mats and logs, and ditches throughout South Carolina.

Eleocharis spp.
Spikerushes
Figures 234-237

All of the 24 species of *Eleocharis* in South Carolina are wetland plants. The spikerushes may be annuals or perennials from stolons or rhizomes. The stems can be 4-angled, 3-angled or terete and are topped by a terminal spike. In several species, the spike is about the same diameter as the supporting stem; otherwise, the spike is distinctly thicker than the stem. The leaves consist of sheaths only, the blades are absent.

The spikes consist of numerous perfect flowers subtended by scales and are usually arranged in a spiral and imbricate manner. The perianth is usually present in the form of 6 downwardly barbed bristles. The achene is lenticular or trigonous and is topped by a cap called a tubercle. Details of the bristles, tubercles, and achenes are important in distinguishing the various species.

The spikerushes are usually found flowering in the summer and early fall. Although a few species frequently grow submersed, most are emergent plants in freshwater and brackish marshes, bogs, flatwoods, and shorelines throughout South Carolina

234
Submersed ***Eleocharis*** *olivacea*

235
Stems of ***Eleocharis microcarpa*** *with terminal spikes*

236
Septate stem and terminal inflorescence of ***Eleocharis equisetoides***

237
Eleocharis quadrangulata *with terminal spike above 4-angled stem*

238
Emergent mat-forming growth

239
Submersed plants in shallow water

240
Small flattened terminal spikes

Eleocharis baldwinii (Torrey) Chapman

Proliferating spikerush

Figures 238-240

This sedge is a stoloniferous mat-forming perennial. The stems are thin, wiry, and 2-15 inches long. Leaf blades are lacking; the leaves consist only of basal sheaths which are often red in color.

The inflorescence is a flattened terminal spike, usually about 1/8 inch long. Each spike contains 1-3 flowers; the scales are reddish-brown and in 2 rows. Flowering occurs from July to September.

A very common aspect of this plant is the vegetative proliferation of plantlets from the spikelet which contributes greatly to the formation of mats. There are other species of spikerush that are similar to *Eleocharis baldwinii* in exhibiting this apical proliferation from the spikelet. These include *E. acicularis*, *E. microcarpa* and *E. vivipara*. The first two species occur throughout the state while the latter is restricted to the coastal plain.

Proliferating spikerush often forms extensive mats of vegetation in marshy areas. It can grow submersed in lakes and ponds, it can grow on or form large floating mats, or it can be found growing as an emergent plant along the margins of lakes and ponds. Proliferating spikerush occurs mainly in the coastal plain of South Carolina.

Scirpus spp.
Bulrushes
Figures 241-243

The bulrushes are annual or perennial wetland and aquatic plants that usually grow in clumps. They can vary in height from just a few inches to over 6 feet. The stems are usually triangular with smooth or sharp angles; however, some species may have terete stems. The leaves consist of blades and sheaths and are basal or on the lower portion of the stem.

The inflorescence is terminal and is subtended by one or more involucral bracts. In those species having only one, the bract looks like a continuation of the stem and the inflorescence appears lateral. The inflorescence consists of ovoid or lanceolate spikelets that may be sessile or in stalked clusters or fascicles. The spikelets are composed of many spirally imbricate scales that subtend the minute flowers. Each flower consists of a pistil, 2 or 3 stamens, and usually 1-6 bristles.

There are at least 16 species of bulrushes occurring in South Carolina that are aquatic or wetland species. They usually are emergent plants; however, at least two species grow as submersed plants. One of these, *Scirpus etuberculatus*, bears some resemblance to *Vallisneria* or to the submersed form of *Sparganium*. It is distinguished from them by its triangular stems and by the branched clusters of ovoid spikelets that are 1/2-1 inch long.

Most bulrush species can be found in flower from mid-summer through mid-fall. Many of the species are limited in their distribution, some are found only in the mountains, others are restricted to the sandhills or to the outer coastal plain. *Scirpus cyperinus*, woolgrass bulrush, is one of the few species found throughout the state. Bulrushes grow in freshwater and brackish marshes, ditches, ponds, streams, swamps and marshy shorelines.

241
Growth aspect of ***Scirpus cyperinus***

242
Young inflorescence of ***Scirpus cyperinus***

243
Submersed ***Scirpus etuberculatus*** *showing strong resemblance to* ***Vallisneria***

Juncaceae

Juncus spp.

Rushes, Bog Rushes

Figures 244-246

244
Juncus effusus *in a wetland habitat*

245
Lateral appearance of inflorescence of ***Juncus effusus***

246
Inconspicuous flowering stems of ***Juncus acuminatus***

Rushes are grass-like, herbaceous, usually perennial, clumping plants. Despite their overall appearance, they are more closely related to the lily family than to the grasses. Over 25 species occur in South Carolina in a wide range of habitats, although the majority are considered to be wetland species.

The stems are usually unbranched with alternate leaves. The leaf blades are either flat or terete and the sheaths are open. A few species produce leaf sheaths only and leaf blades are not formed.

The inflorescence is terminal on the stem and is subtended by bracts. In one group of rushes, the main involucral bract looks like a continuation of the stem thus giving the impression that the inflorescence is lateral. The flowers are in cymes or clusters, often appearing head-like. Each flower has 3 sepals and 3 petals which are all bract-like or scale-like and usually green, brown or dark red in color. The fruit is a capsule containing numerous tiny seeds.

Juncus species usually flower from late spring or early summer through the fall. Some species are quite restricted, growing only on granite outcrops or other specialized habitats. However, many species of *Juncus* will grow anywhere there is moist soil or shallow water.

Juncus repens Michaux
Creeping Rush
Figures 247-249

Creeping rush is a plant that exhibits different growth forms in different habitats. When growing on moist soil, it is a short, tufted plant from 2-8 inches tall. When it occurs as a submersed plant, the stems become lax, elongate and root at the nodes. Submersed plants are usually bright green in color and can form large colonies. Submersed leaves are linear, flaccid, and fascicled or subopposite at the nodes.

Flowers are produced in compact heads of 3-15 flowers each. Individual flowers are green, usually about $^1/_3$ inch long, with rigid, narrow sepals and petals. The fruit is a many-seeded capsule about $^1/_4$ inch long.

Emersed and terrestrial plants flower during the summer and fall. Submersed plants usually do not flower unless water levels drop exposing the plants. *Juncus repens* is typically found in ponds, lakes, and ditches in the sandhills and coastal plain.

247
Short, tufted emergent form

248
Elongate, flaccid submersed form

249
Linear leaves fascicled at nodes

V. Algae

Contributed by:
Lawrence A. Dyck

Introduction to Algae

This chapter is a guide to the identification of nuisance algae commonly encountered in South Carolina. Not included in this section are the several hundred species of microscopic algae that form the phytoplankton and periphyton component of the food chain in rivers, ponds and reservoirs of the Southeastern United States. Instead, ten common taxa are presented to illustrate three nuisance types: (1) plankton blooms caused by microalgae; (2) infestations caused by filamentous and colonial macroalgae; and (3) submersed algae resembling vascular plants.

The intent is to emphasize the field characteristics of algae that can be detected through the observer's sight, feel and smell. However, it is important to understand that critical identifications, even to the level of genus used in this manual, require microscopic inspection.

Two reference works that provide supplemental information are: *Freshwater Algae of the United States* by G. M. Smith, and *A Manual of Fresh Water Algae* by L. A. Whitford and G. J. Schumacher. The first named reference describes many genera found throughout the United States, the second reference provides keys to many of the algal taxa in the Carolinas.

Planktonic Algae

Euglenoid Algae

Euglena Ehrenberg

Figures 250-252

The single-celled organisms classified within the genus *Euglena* (the "Good Eyeball Algae") are world-wide in distribution and belong to more than 100 species. Those species of *Euglena* that cause nuisance blooms do so in environments enriched with organic materials, usually animal wastes. Blooms take the form of a surface scum or film that may vary in color from lime- or yellow-green to blood-red. Surface-films typically vary in magnitude throughout the day, becoming particularly noticeable in sheltered areas during the early morning hours and diminishing as the sun's intensity increases. Surface variations occur as the flagellated cells migrate within the water column, moving to deeper layers during the day and returning to the surface at night. If a surface-film is penetrated by an object (e.g., canoe paddle) it produces characteristic "cracks" which do not reunite readily.

Microscopic examination of surface films is required for visualization of the planktonic cells. The motile, fusiform-shaped cells are pulled through the water in a spiral fashion by a single anterior flagellum that beats with "lasso-like" rotations; in addition, cells undulate by movements of the cell-body itself. Cells have a distinct eyespot located adjacent to the flagellar base. Euglena reproduces by cell fission and exists throughout all seasons in its vegetative form.

Euglena-blooms are particularly common in farm and aquaculture ponds that receive a high influx of organic material. Blooms occur throughout the year, but peak in the warmer months of the year.

250
Aspect of red-colored surface-bloom

251
Surface-bloom showing "cracks" in the film

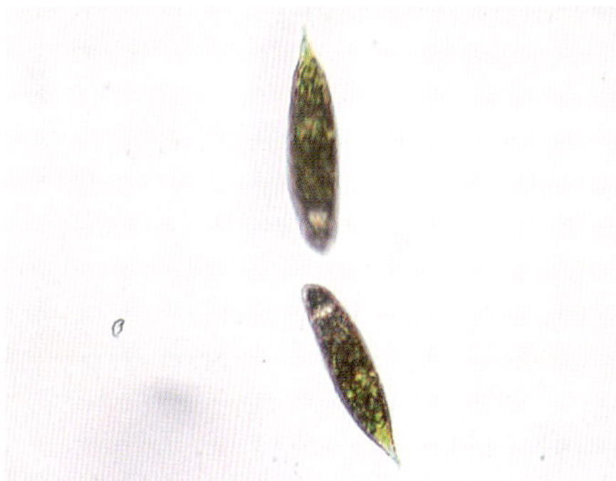

252
Microscopic view of fusiform cells

Blue-Green Algae/ Cyanobacteria

Microcystis Kuetzing and *Anabaena* Bory

Figures 253-256

253
Aspect of uniformly suspended blue-green algae

254
Aspect of surface-bloom

Members of the genera *Microcystis* ("Small Celled Alga") and *Anabaena* ("Alga that Floats Up") are commonly associated with blooms of planktonic blue-green algae (also called cyanobacteria). Blooms have several characteristics that assist with their identification. From a distance, blooms of planktonic cyanobacteria give the water a uniform blue-green to lime-green coloration. On closer inspection, however, the water will appear filled with small flakes of darker blue- or lime-green material, and in sheltered areas these small flakes will aggregate into extensive surface scums, sometimes covering an entire cove with a surface layer up to 3 inches thick.

Surface blooms or scums usually form over many weeks, and the particulate "flake-stage" is detectable well in advance of the surface phenomenon. Flakes may vary in size from barely detectable specks, to clumps several inches in diameter. Often flakes give the appearance of suspended grass-clippings.

An individual flake is an aggregation of planktonic blue-green algae. Through interaction with diel circulation patterns, these planktonic aggregates regulate their vertical migrations on a daily basis. At night, the planktonic cells synthesize air-filled gas-vacuoles which buoy the flakes to the surface; whereas during daylight, intense photosynthetic activity results in osmotic collapse of gas-vacuoles and flakes sink to lower depths. Under normal climatic conditions the cycle of rising and falling continues. However, during calm weather flakes may rise and aggregate into ever-larger particles, until the enlarged-aggregates are trapped at the surface by their expanded surface area. At this point a surface bloom has formed and will persist for extended periods. Dispersal of the surface-bloom is not well understood, but involves storms in combination with phenomena that interfere with the alga's metabolic-health.

Microcystis flakes are comprised of hundreds to tens-of-thousands of tiny cells (ca. 5 µm in diameter) which aggregate within a common mucilaginous matrix. *Anabaena* flakes are made of unbranched, coiled filaments comprised of spherically-shaped cells (ca. 10 µm in diameter) all encased within a common mucilage. *Anabaena*, unlike *Microcystis*, may contain specialized heterocysts that permit nitrogen fixation. Typically, cyanobacterial cells show little internal organization; however, when gas-vacuoles are present, they appear as clear areas within the diffusely colored cytoplasm.

Cyanobacterial blooms are prevalent in the warm waters of late summer and early fall in those lakes and ponds subject to high nutrient input (particularly elevated phosphate). These blooms often follow the decline in planktonic communities dominated by green algae.

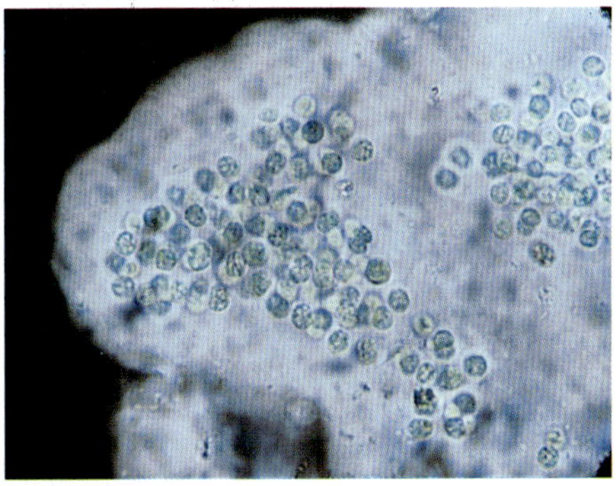

255
Microscopic view of ***Microcystis****: note the many small cells in a common mucilage*

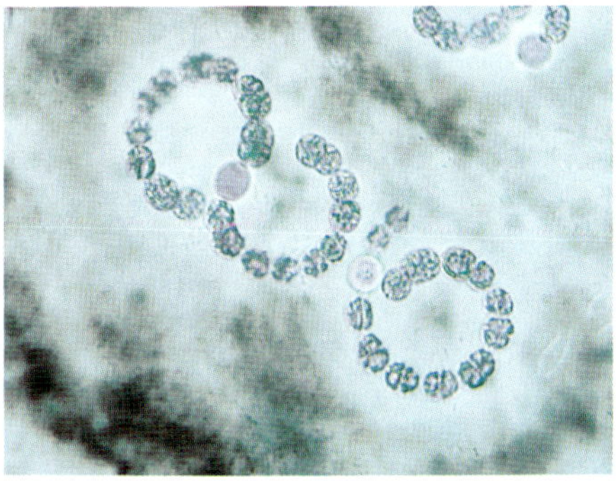

256
Microscopic view of ***Anabaena****: note intercalary heterocyst and gas-vacuole containing vegetative cells*

Green Algae

Scenedesmus Meyen

Figures 257-258

257
Aspect of a pond with a green algal bloom

258
Microscopic view of colonies

Blooms of green algae cause ponds to appear bright-green to olive-green. The color is generally dispersed uniformly and surface films or scums are rare. Blooms are most common in late spring to early summer and are associated with nutrient enrichment, usually nitrate and phosphate.

Many different green algae are responsible for the coloring of waters associated with green-algal blooms. The example used here is a common colonial organism consisting of 2, 4 or 8 (rarely more) cells joined laterally into flat groups. The terminal cells of the colony usually produce spines that assist in flotation. *Scenedesmus* is a common member of the plankton but shares the habitat with many other motile and non-motile unicellular and colonial green organisms.

Green algal blooms are often favored for use as fish food and are encouraged through heavy fertilization.

Filamentous and Colonial Algae

Blue-Green Algae/Cyanobacteria

Lyngbya Agardh

"Musty Black Mat Alga"

Figures 259-262

The genus *Lyngbya* can be separated into two groups: (1) taxa with small filaments that form thin membranous mats; and (2) taxa with large coarse filaments that entangle into extensive "woven" mats. It is the latter group which forms nuisance infestations and is described and illustrated here.

During warmer months, *Lyngbya* mats appear as either floating-islands or as broad expanses that cover entire ponds and coves. Typically, mats are uniformly black in the spring, but become mottled with patches of black, brown, green and white filaments in late summer and fall. Trailing from the surface mats are long "streamers" which may hang several feet into the water column and are physiologically the most active portion of the mat. In addition, extensive benthic or bottom mats also exist and persist throughout all seasons. Portions of benthic mats rise to the surface when oxygen becomes trapped and buoys the mat to the surface. Similarly, surface mats fall back to the bottom when oxygen is forced out of mats by thundershowers or turbulence.

Mats are comprised of highly entangled filaments with the texture and appearance of long tangled strands of coarse human hair. They have a strong distinctive musty or earthy odor. A hand-held mat will retain its entangled integrity while draping in response to gravity.

Microscopically, filaments are about 50 µm in diameter and are comprised of stacks of disc-shaped cells with homogeneous, organelle-free cytoplasm colored green, purple or black. Cells are confined within a distinct clear sheath. Empty sheaths persist within the entangled mat. On occasion, false branches occur when a stack of cells grows through the side of a sheath.

Infestations are associated with elevated calcium levels and may also be related to increased nutrient loads (primarily phosphorous).

259
Aspect of heavy surface-infestations

260
Aspect of mottled mats

261
Hand-held mat

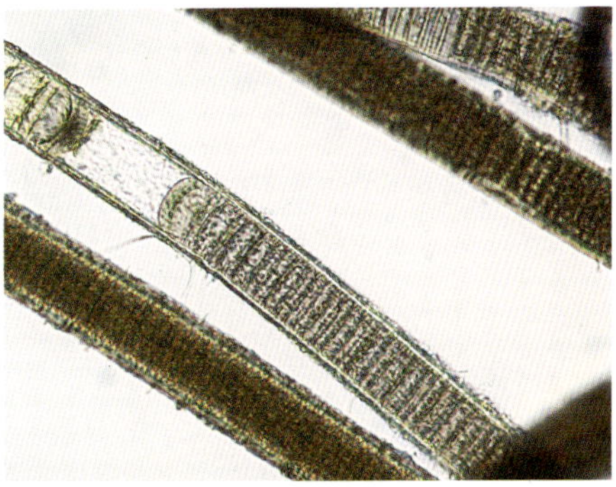

262
Microscopic view of filaments: note disc-shaped cells surrounded by a clear sheath

Green Algae

Hydrodictyon Roth

Water Net

Figures 263-266

263
Aspect of infested cove

264
Close-up showing frothy nature of surface aggregations

265
Several net-colonies adhering to a canoe paddle

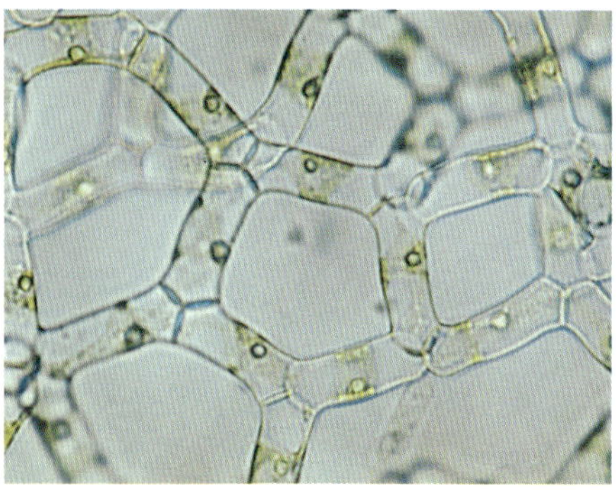

266
Microscopic view of cells in a net-colony

In its infestation condition, this alga forms extensive frothy surface-aggregations. The surface-aggregations are medium to pale green in color and are formed when a multitude of cylindrically shaped net-like colonies rise to the water's surface. When removed from the surface by hand, the aggregation of nets readily dissociates and the mass falls apart. Individual colonies are formed as hollow cylinders with a perimeter of cells forming a mesh-like network. When first formed, cylinders are minute, having arisen as the reorganized contents of a single cell within a parent-net. However, with time each cell of the colony expands, and the entire colony attains a length of 6-8 inches and a width of 1-2 inches. Inspection of the infested area will show net-colonies aggregated at the water's surface, entangled with submerged vegetation and resting on the bottom.

Microscopic examination reveals the cellular nature of mesh-like networks: the perimeter of each colonial-cylinder is comprised of cylindrical cells arranged in groups of 5 or 6 to form the pentagonal and hexagonal compartments of the net. As nets enlarge, their cell number remains constant while their individual cells enlarge many-fold.

Infestations occur in mid-summer and last for several weeks, after which the nets disappear, having shed perpetuating spores and zygotes into the sediments. Infestations are usually confined to hard-water environments that are relatively protected and have elevated nutrient levels.

Oedogonium Link
"Entangled Green Filaments with Microscopic Growth-Rings"
Figures 267-270

This alga grows attached to substrates, usually submerged logs, branches, leaves, aquatic animals or other algae. In addition, the alga attaches to rocks or the walls of concrete drainage canals where extensive growth may impede water-flow. The alga is comprised of unbranched filaments that collectively impart a dark-green to yellow-green color. This alga is most readily confused with *Spirogyra* in the field. However, the two genera are distinguished by touch and by growth habit. *Spirogyra* is slimy, and its filaments readily slide from one's hand, whereas *Oedogonium* will feel slimy when rubbed, but its filaments adhere to the hand. In growth habit, *Spirogyra* filaments grow parallel and close to one another, and while *Oedogonium* filaments may be aggregated, the strands overlap one another and give a more tangled appearance.

Diagnostic microscopic features include: unbranched filaments comprised of cylindrical to barrel shaped cells with a reticular chloroplast and a unique mechanism of cell expansion that leaves "expansion-rings" at one end of a parent cell following each episode of growth. In addition, sexual reproduction provides distinctive dark, spherical oogonia/oospores intercalated among cylindrical vegetative cells.

Oedogonium is a cosmopolitan genus. It grows throughout the year in freshwater of all pH values.

267
Aspect of detached mats

268
Hand-held alga: note that it does not slide off hand

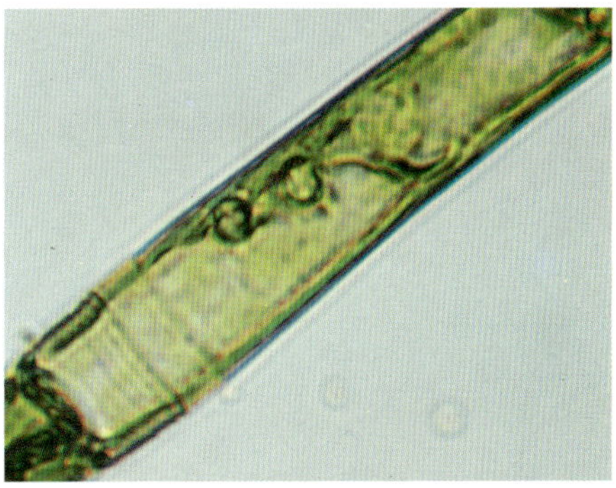

269
Microscopic view of expansion rings

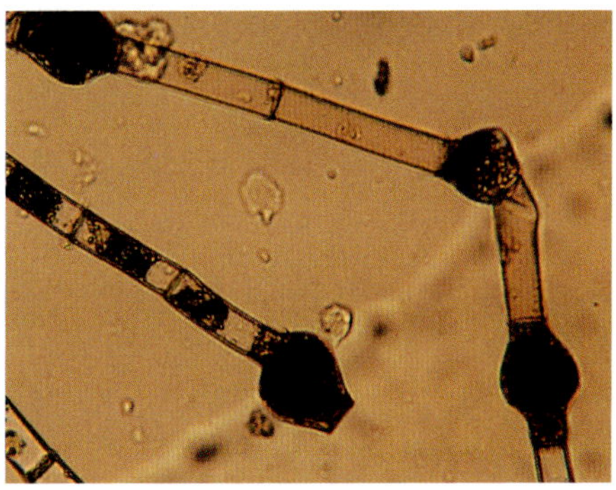

270
Microscopic view showing spherical oogonia

Pithophora Wittrock

"Green Cotton-ball Alga"

Figures 271-274

271
Aggregation of floating cotton-ball mats

272
Coarse character of cotton-ball masses

273
Hand-held cotton-like mass

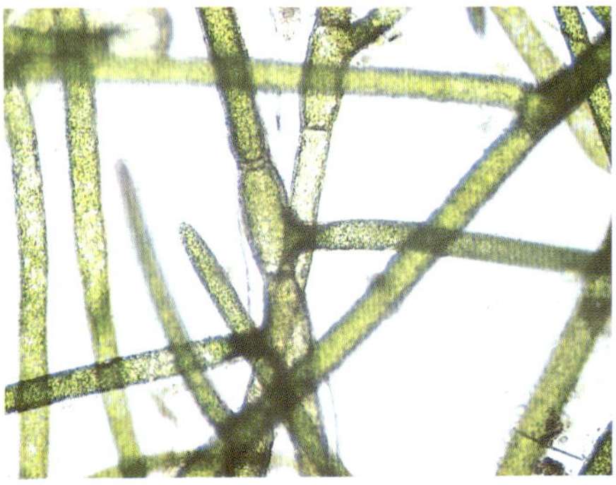

274
Microscopic view of cylindrical cells, branches and barrel-shaped akinetes

Pithophora forms major surface infestations, often covering entire coves and shallow ponds. The alga consists of aggregated, floating green masses which vary in diameter from the size of a quarter coin to that of dinner plate. Cotton-like masses vary in color from dark green to lime green, and older regions, at the base of a mass, may be encrusted with iron deposits to give a rust or black coloration.

The cotton-like masses are coarse to the touch and lack the slimy texture of many other filamentous algae. Each cottony-mass is easily picked up and is difficult to tear apart. When hand-held, the coarse branched character of filaments can be distinguished with the naked eye, and the mass of branched material does not collapse, but remains tuft-like. These characteristics contrast with *Lyngbya*, an alga that forms extensive surface mats similar to *Pithophora*, but is unbranched and will collapse when hand-held.

Microscopically, *Pithophora* has filaments of long, multi-nucleate, cylindrical cells containing many peripheral green chloroplasts. Branches are common, and arise from axial-cells, just below the apical cross wall. Filaments usually contain an abundance of dark green or brown colored akinetes. Akinetes are resistant, spore-like cells that overwinter in the sediments and germinate to produce new cottony-tufts the following spring.

Infestations are most common in the warmer months of the year, but in heavily infested ponds they persist year-round. Infestations are associated with hard-waters subject to nitrogen and phosphorous enrichment.

Spirogyra Link
"Green Silky-Strand Alga"
Figures 275-278

This common alga is found on the bottom, floating as small to moderate sized island-mats and entangled with logs, emergent vegetation and other floating materials. Its color is bright green in the spring, but may bleach to a yellow-green later in the season, or as filaments degrade. On close inspection in the field, the unbranched filaments will be oriented as closely-parallel strands. When lifted from the water, the silky quality of filaments is easily detected. Filaments will readily slide between one's fingers and fall from a grasp.

Microscopic inspection of the filaments reveals spiral-shaped chloroplasts which lie at the periphery of all cells and coil uniformly along each cell's length. Other diagnostic characteristics include a two layered cell wall (the outer layer has the slimy character) and a unique mechanism of reproduction by conjugation tubes that results in the production of dark colored zygotes. Fragmentation of filaments and germination of zygotes are the only known mechanisms for reproduction.

In South Carolina, *Spirogyra* is found throughout the year in both hard and soft waters. Its extensive growth in the spring may produce bright green mats that clog seepage areas, drainage canals, and even boat canals and coves.

275
Aspect of an infested seepage area

276
Detection of silky quality of filaments

277
Microscopic view showing spiralled chloroplasts of two species

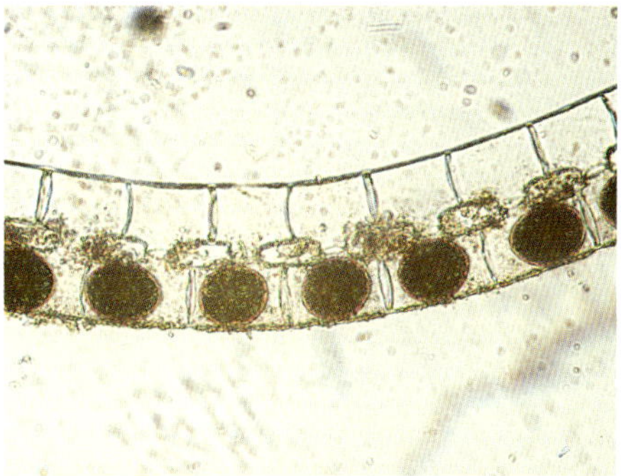

278
Microscopic view of conjugation tubes and zygotes

Yellow-Green Algae

Vaucheria De Candolle

"Attached Felt-Like Alga"

Figures 279-280

279
Aspect view of benthic masses

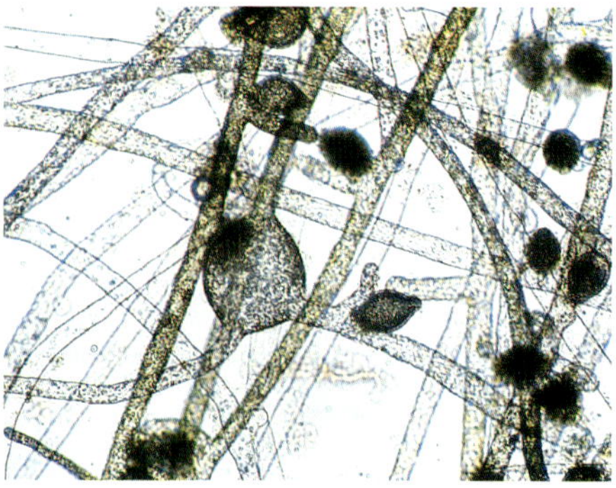

280
Microscopic view of tubular coenocytes, several dark ovoid oogonia and coiled antheridia

This is a benthic alga that grows attached to the substrate in shallow areas which may become terrestrial. These felt-like masses may expand so as to choke shallow tributaries, seepage areas, and runoff canals, and thus contribute to flooding in backwater areas.

Vaucheria is light green, grey-green or dark-green to black in color and is soft to the touch, but not slimy. The alga anchors itself into the substratum by rhizoids and tightly interweaves its photosynthetic filaments. Hence, its removal from the substrate is usually as a mass of material.

Microscopic examination reveals that the filaments are extended tubes without cross-walls (coenocytes) except where reproductive structures are formed. The cytoplasm is characterized by numerous ovate chloroplasts along the periphery that migrate with the cytoplasm and may be extruded when the coenocyte is broken. Branches arise laterally or dichotomously and are the same size as the main filament. A sexual reproduction occurs through multiflagellate zoospores, and sexual reproduction involves oogonia containing solitary eggs and antheridia that produce many antherozoids.

Macroalgae
Charophyta
Chara Linnaeus and *Nitella* Agardh
Musk-grass or Stonewort

Figures 281-283

Chara forms extensive submersed beds with a structure that mimics the habit of vascular aquatic plants. Individual plants range from a few inches to several feet in height, depending upon the species. A series of fine colorless rhizoids that anchor the alga firmly within the substrate. The plants often feel rough to the touch due to calcium carbonate deposition and typically have a garlic-like musty odor.

Chara has whorls of 6-8 branchlets that radiate from nodes and alternate with elongated internodes. Each internode is formed by a single "giant cell" up to 3-4 inches long, which, depending upon species, may or may not be surrounded by a cortex of vertically oriented smaller cells. Both monoecious and dioecious species exist, and reproduction occurs through specialized multicellular sex structures located on the branchlets of upper whorls. Here, bright orange colored antheridia produce motile sperm in summer and fall and fertilize white to light orange colored eggs (oogonia). Upon fertilization, oogonia become black colored oospores about the size of small strawberry "seeds". Oospores serve as propagules and may germinate into new individuals after confinement to the sediments for several seasons. In most regions of South Carolina, however, oospore germination may be of limited importance since vegetative plants persist throughout the year. *Chara* favors sheltered environments with sandy sediments of low organic content that are bathed by slow moving hardwater with good clarity.

Nitella, a close relative of *Chara*, grows in meadows within somewhat deeper water. *Nitella* is distinguished from *Chara* based upon the branching and bushy character of nodal branchlets and by the absence of either lime encrustations or the musty or garlic odor.

281
Aspect view of a ***Chara****-bed*

282
Close-up of ***Chara****, note that branchlets do not have branches*

283
Close-up of ***Nitella****, note that branchlets have branches*

Glossary

Achene–a hard, dry, one-seeded fruit.

Acuminate–tapering to a point such that the tapering margins are concave. (Plate V-A).

Acute - tapering to a point such that the tapering margins are straight to convex. (Plates III-E, VI-C).

Alternate–referring to the arrangement of plant parts, usually leaves; leaves are alternate when they appear one per stem node. (Plate I-A, III-G, IV-A).

Annual–plants that survive the winter as seeds, completing their life cycle in a single year.

Anther–site of pollen production in the stamen. (Plate VI-A).

Awn–a needlelike appendage.

Axil–the upper angle that a leaf makes with its associated stem; the bud found in that angle is an axillary bud. (Plates I-B, III-B, IV-A, VI-B).

Blade–the flat, expanded portion of a leaf. (Plates III-C, IV-G).

Bract–a modified leaf usually developing below a flower or floral cluster; this modified leaf or bract may be scale-like as in sedges and grasses. (Plate I-C).

Calyx–the outermost whorl of the flower; these floral parts are usually green and are individually known as sepals. (Plates I-C, II-A, IV-E, IV-F).

Capillary–slender and threadlike. (Plate I-D).

Capsule–a dry, dehiscent fruit that breaks open along two or more sutures or lines to expose seeds as in primrose. (Plates I-B, II-A).

Ciliate–referring to a flattened structure (usually a leaf, petal or fruit) bearing hairs only on the margins.

Claw–a narrow constricted base to some sepals and petals as in the petals of *Ludwigia leptocarpa* or *Decodon verticillatus*. (Plate IV-E, IV-F).

Compound–consisting of two or more similar parts, as in leaflets to form a compound leaf; referring to a branched inflorescence with stalked or pedicelled flowers. (Plate I-E).

Cordate–heart-shaped; referring to a base with rounded, shallow lobes. (Plates I-F, III-F, IV).

Corm–a scaled, bulbous structure at the base of some plants as in *Sagittaria*. (Plate IV-G).

Corolla–the showy, often brightly colored portion of a flower; the individual units of this floral whorl are petals. (Plates I-C, I-G, IV-E, IV-F).

Crenate–having a scalloped margin with shallow rounded teeth as in the leaf margins of *Hydrocotyle umbellata*. (Plate III-H).

Deciduous–being not evergreen; usually refers to all leaves lost at the end of the growing season.

Decumbent–lying or growing along the ground but with the tips turning upward. (Plate I-H).

Decurrent–attached to a stem or petiole and extending as a wing beyond that point of attachment. (Plate I-I).

Dehiscent–splitting open at maturity as in the capsule of *Hibiscus*. (Plate I-B, II-A).

Dentate referring to a margin with sharp, outward-pointing teeth lying at right angles to the margin.

Dichotomous–forked or branched into equal pairs. (Plates II-B, VI-G).

Dimorphic–having two forms as in the differing shapes of emersed and submersed leaves of *Myriophyllum heterophyllum*.

Dioecious–possessing imperfect (staminate or pistillate) flowers on separate plants of the same species. (Plate VI-B).

Dissected–having a margin deeply divided into numerous, often irregular, segments. (Plates I-D, II-B, II-C).

Distichous–referring to similar structures being two-ranked and lying in the same plane relative to one another as in the leaf arrangement of *Xyris*. (Plate II-D).

Elliptic–having a two-dimensional shape of an ellipse, being widest at the structure's midpoint. (Plate III-C, VI-C).

Emergent–rising out of and above the surface of the water; emersed. (Plate IV-G).

Emersed–emergent, normally found above the surface of the water.

Entire–having a smooth margin without indentations or incisions. (Plate I-F, II-F, III-C).

Fibrous–having loose, woody fibers; often referring to root systems that are extensively branched in a fibrous network. (Plate VI-D).

Filament–a threadlike structure; the stalk below the anther in most stamens. (Plate VI-A).

Filiform–threadlike; having long slender filaments. (Plate II-C).

Flower–complete flowers are composed of sepals, petals, stamens and pistil(s); incomplete flowers lack one or more of these floral whorls. (Plates I-C, IV-H, V-H, VI-B).

Glabrous–lacking hairs; not pubescent.

Glaucous–having a whitened or pale appearance due to a powdery or waxy covering.

Globose–shaped like a globe or sphere. (Plate II-E).

Hastate–arrow-shaped with pointed, often narrow, bases in an abrupt outward orientation.

Head–a dense cluster of sessile or subsessile flowers; a capitulum with sessile flowers borne on a compound receptacle. (Plate II-F).

Herb–plant lacking persistent woodiness, being herbaceous, aerial plant parts usually not overwintering.

Imbricate–refers to structures regularly overlapping by their margins as in the imbricate leaves of *Azolla*. (Plate II-G).

Imperfect flower–a unisexual flower lacking either pistils (staminate) or stamens (pistillate). (Plates V-F, V-G, VI-B).

Indehiscent–not splitting open at maturity as in the fruit of *Nelumbo*. (Plate II-H).

Inflorescence–an arrangement of one or more flowers on a floral axis. (Plate II-F, IV-H, V-G, V-H).

Internode–region between nodes on a stem where no leaves or branches originate. (Plates I-H, II-I, III-G).

Keeled–having a longitudinal ridge resembling the keel of a ship; referring to a ridge as in the low-keeled nerve of a *Hydrilla* leaf. (Plate V-D).

Lanceolate–lance-shaped; having a shape much longer than wide, being widest just above the base and gradually tapering to the tip. (Plate IV-A).

Leaflet–a smaller leaf-like division of a compound leaf as in *Cicuta*. (Plate I-E).

Lenticular–biconvex or lens-shaped as in the lenticular leaves of the duckweed family, Lemnaceae. (Plate III-A).

Linear–having a long, narrow shape with evenly wide sides. (Plates II-B, II-D, I-C).

Lobed–having rounded segments along the margins. (Plates I-F, IV-G, V-C).

Midrib–the central vein of a leaf or flattened structure. (Plates I-A, II-F, V-D, VI-C).

Monoecious–possessing imperfect (staminate and pistillate) flowers on one plant as in *Limnobium* or *Peltandra*. (Plates V-F, V-G).

Mucilaginous–being gummy or gelatinous as seen in the mucilaginous coating of submersed portions of water-shield (*Brasenia*) leaves. (Plate III-C, IV-D).

Node–stem region or joint from which a leaf or leaves, branch or branches, originate. (Plates II-I, III-G, V-A).

Nodose–having a nodular or knotty stem as in *Polygonum* or *Potamogeton nodosus*. (Plates III-D, III-G).

Nutlet–a small nut.

Oblanceolate–inversely lanceolate; having a shape much longer than wide, being widest above the middle and gradually tapering to the tip as in *Alternanthera* leaves. (Plate II-F).

Oblong–having a longer than wide shape with nearly parallel margins as in *Hydrilla* leaves. (Plate V-D).

Obovate–inversely ovate; inversely egg-shaped, being widest above the middle as in the leaves of *Spirodela polyrhiza* or the petals of *Ludwigia* or *Decodon*. (Plates IV-E, IV-F).

Obtuse–rounded or bluntly angled at the tip as in the leaves of *Echinodorus cordifolius* . (Plate III-F).

Ocrea–a stipular sheath at a stem node characteristic of the smartweed family, Polygonaceae. (Plates III-D, III-G).

Opposite–referring to the arrangement of plant parts, usually leaves; leaves are opposite when they appear two per stem node. (Plates I-D, I-H, II-E, II-F).

Orbicular–having a circular outline as in the leaf shape of *Hydrocotyle umbellata*. (Plates II-G, III-H).

Ovary–basal portion of a pistil which contains one or more ovules; matures into a fruit with one or more seeds. (Plate IV-A).

Ovate–egg-shaped, being widest below the middle. (Plate III-F).

Palmate–having three or more veins, lobes or leaflets arising from one central point to a hand- or palm-like fashion as in the leaf blade venation of *Sagittaria*. (Plate IV-G).

Panicle–a loose and irregular branching cluster of flowers; a compound inflorescence with pedicellate or stalked flowers.

Papilla–a warty or short, blunt projection as seen on the watermeal *Wolffia*. (Plate IV-B).

Pedicel–the stalk or stem of an individual flower in an inflorescence. (Plates IV-C, IV-H).

Pedicellate–having a stalk or pedicel. (Plates IV-C, IV-H).

Peduncle–the stalk of a solitary flower; the main stalk of an inflorescence to which the pedicels of individual flowers are attached. (Plate IV-C).

Peltate–referring to the point of attachment of a plane structure such as a leaf blade; the attachment is at a point not at the margins; referring to the central location of petioles such as in water lotus or marsh pennywort leaves. (Plates III-C, III-H, IV-D).

Perennial–living more than two years.

Perfect flower–a bisexual flower having both pistil(s) and stamens. (Plates I-G, IV-E, IV-F, V-H).

Perianth–a collective term for the sepals (calyx) and petals (corolla) of a flower. (Plates I-C, IV-E, IV-F).

Petal–the showy, often brightly colored part of a flower; a unit of the corolla whorl of a flower. (Plates IV-E, IV-F, VI-B).

Petiole–the stalk of a leaf. (Plates I-E, III-H, IV-G, IV-I).

Pinnate–referring to a feather-like arrangement of parts along a common axis or rachis; referring to the even pattern of divergence of branches, veins or leaf divisions on either side of an elongate axis; in compound leaves, having the leaflets attached at more than one point to the petiole or rachis. (Plates I-A, III-G, IV-A).

Pistil–the innermost structure or whorl in a flower; its swollen base is the ovary, the site of ovule(s) or seed(s).

Pistillate–referring to a unisexual flower containing pistil(s), but lacking stamen(s); an imperfect flower.

Procumbent–lying or trailing along the ground but not rooting at the nodes.

Pubescent–covered with hairs.

Punctate–covered with small glandular or colored dots or pits as on the leaf surfaces of *Bacopa* or *Wolffia*. (Plate IV-B).

Quadrangular–square in outline. (Plate I-I).

Raceme–a simple elongate inflorescence with stalked or pedicellate flowers as in water-spider orchid (*Habenaria*), or lizard's tail (*Saururus*) inflorescences. (Plate IV-H).

Rachis–a central axis or stem bearing either flowers or leaflets. (Plate I-E).

Receptacle–the expanded tip of a stem or pedicel that bears one or more flowers. (Plate II-H).

Recurved–curved backward or downward. (Plate VI-E).

Reniform–kidney-shaped as in floating-heart (*Nymphoides*) or frog's bit (*Limnobium*) leaves. (Plates I-F, IV-I).

Repent–lying or trailing along the ground; rooting at stem nodes. See procumbent. (Plate V-A).

Reticulate–netted.

Rhizoid–a root-like structure.

Rhizome–a horizontal underground stem distinguished from a root by having nodes. (Plate II-I).

Rosette–a cluster of structures radiating from a central point of attachment as in the floating rosette leaves of frog's bit, *Limnobium spongia.* (Plate V-B).

Sagittate–arrow-shaped with basal lobes pointing downward as in the leaves of arrowhead, *Sagittaria latifolia.* (Plates IV-G, V-C).

Scabrous–rough to the touch.

Scape–a leafless flowering stem abruptly arising from a rosette of leaves or from below the substrate but not from other evident stems; scapes occur in bladderwort (*Utricularia*) and in creeping burhead (*Echinodorus cordifolius*). (Plate V-A).

Sepal–one of the individual floral units of the calyx; usually green. (Plates II-A, IV-A, IV-E, IV-F).

Septate–partitioned into evident segments or chambers.

Serrate–referring to a margin with sharp, ascending teeth. (Plates I-E, VI-G).

Serrulate–finely serrate along the margins as in the serrulate leaves of Brazilian elodea, *Egeria densa.* (Plates III-E, VI-B).

Sessile–lacking a stalk or petiole at a structure's point of attachment as in the leaves of tape grass (*Vallisneria*) or alligatorweed (*Alternanthera*). (Plates I-H, II-E, V-D).

Sheath–a tubular enveloping structure; the tubular base of a leaf surrounds the stem as in the sheathing petioles of lizard's tail leaves (*Saururus*). (Plates I-E, II-I, III-B, V-E).

Simple–a leaf with a single blade; an unbranched inflorescence. (Plates I-A, V-H).

Sinus–the recess between two segments or divisions as in the space between the basal lobes of an arrowhead leaf (*Sagittaria*). (Plates I-F, IV-F, V-C).

Spadix–a spike with flowers sunken in its fleshy rachis, as in the inflorescence of the arrow arum, *Peltandra virginica.* (Plates V-F, V-G).

Spathe–a sheath that envelops a flower or inflorescence; often associated with a spadix. (Plates V-F, V-G, VI-B).

Spatulate–spoonlike; being widest near a rounded tip as in the spatulate petals of *Hydrilla.*

Spike–a simple elongate inflorescence with flowers as in the spikes of pondweeds, *Potamogeton*, or pickerelweed, *Pontederia.* (Plate V-H).

Spikelet–a small spike of sessile flowers with these flowers lacking a perianth, being only bracted, and considerably reduced in size; characteristic of grasses and sedges.

Spinulose–having minute spines or stiff points along the margins as in the spinulose margins of naiad or hydrilla leaves. (Plates III-B, V-D).

Spur–an elongate tubular extension of a petal or sepal such as seen on bladderwort (*Utricularia*) or fringed orchid (*Habenaria*) flowers; a shortened branch with crowded leaves. (Plate V- I).

Stamen–part of a flower, usually consisting of a filament and an anther, site of pollen production. (Plate VI-A, IV-F).

Staminate–referring to a unisexual flower containing stamens but no pistil; an imperfect flower. (Plates V-F, V-G, VI-B).

Stipule–a paired structure occurring at leaf bases, may be leafy, glandular, scaly, spiny or fused together, often inconspicuous or even absent but may reach a length of three inches or more in the pondweed, *Potamogeton nodosus.* (Plate VI-C).

Stolon–a horizontal runner; a specialized stem with long internodes and capable of rooting at the nodes to establish new plantlets; means of vegetative propagation as seen in tape grass (*Vallisneria*) or frog's bit (*Limnobium*). (Plate VI-D).

Stoloniferous–producing stolons. (Plate VI-D).

Submerged–submersed. (Plates II-C, III-E).

Submersed–normally growing underwater; a plant may grow totally submersed or may also have some parts emersed. (Plates II-B, III-C).

Subtend–to be closely associated with another structure and developing from a point beneath it; a subtending structure may also enclose another as in *Hydrilla* or *Egeria* spathes subtending the flowers. (Plates I-C, V-G, VI-B).

Taproot–a persistent primary root that is well-developed and dominant.

Terete–round.

Toothed–having sharp teeth along the margins as in hydrilla or naiad leaves. (Plates III-B, V-D).

Truncate–referring to a tip or base ended squarely across a structure as if cut.

Tuber–a fleshy enlarged rhizome or stolon. (Plate IV-I, VI-F).

Turion–a scaly, thickened winter bud produced at leaf axils in aquatic plants such as hydrilla enabling these plants to disperse. (Plate VI-E).

Umbel–a flat-topped inflorescence of stalked or pedicellate flowers with all the pedicels arising from a single central point; an inflorescence characteristic of the carrot family and seen in pennywort, *Hydrocotyle*. (Plate IV-C).

Undulate–referring to a wavy margin such as seen in water lotus leaves, *Nelumbo*.

Vein–conducting channels in a plant responsible for the movement of water, food and minerals; veins are obvious as lines on the surface of leaves.

Venation–the pattern of veins of a leaf. (Plates I-A, II-D, IV-G).

Whorl–a cyclic arrangement of three or more structures as exemplified by the whorled leaves of the coontail, *Ceratophyllum*. (Plates II-B, II-C, III-B, III-E, V-D).

Wing–a flange of tissue as on the winged stems of *Ludwigia decurrens*. (Plate I-I).

PLATE I.

A. *Nyssa aquatica* leaves in alternate position with pinnate venation.
B. *Hibiscus moscheutos* dehiscent capsule in leaf axil.
C. *Hibiscus moscheutos* perianth of corolla and calyx with calyx subtended by linear bracts.
D. *Cabomba caroliniana* leaf with capillary dissected segments and opposite position to second leaf.
E. *Cicuta maculata* compound leaf with leaflets attached to rachis above a sheathing petiole base, leaflets with serrate margins.
F. *Nymphoides* leaf blades with cordate rounded bases and entire margins.
G. *Echinodorus cordifolius* perfect flowers, one in bloom with expanded corolla.
H. *Alternanthera philoxeroides* decumbent stem and opposite sessile leaves.
I. *Ludwigia decurrens* leaf blade decurrent onto wing-angled (quadrangular) stem.

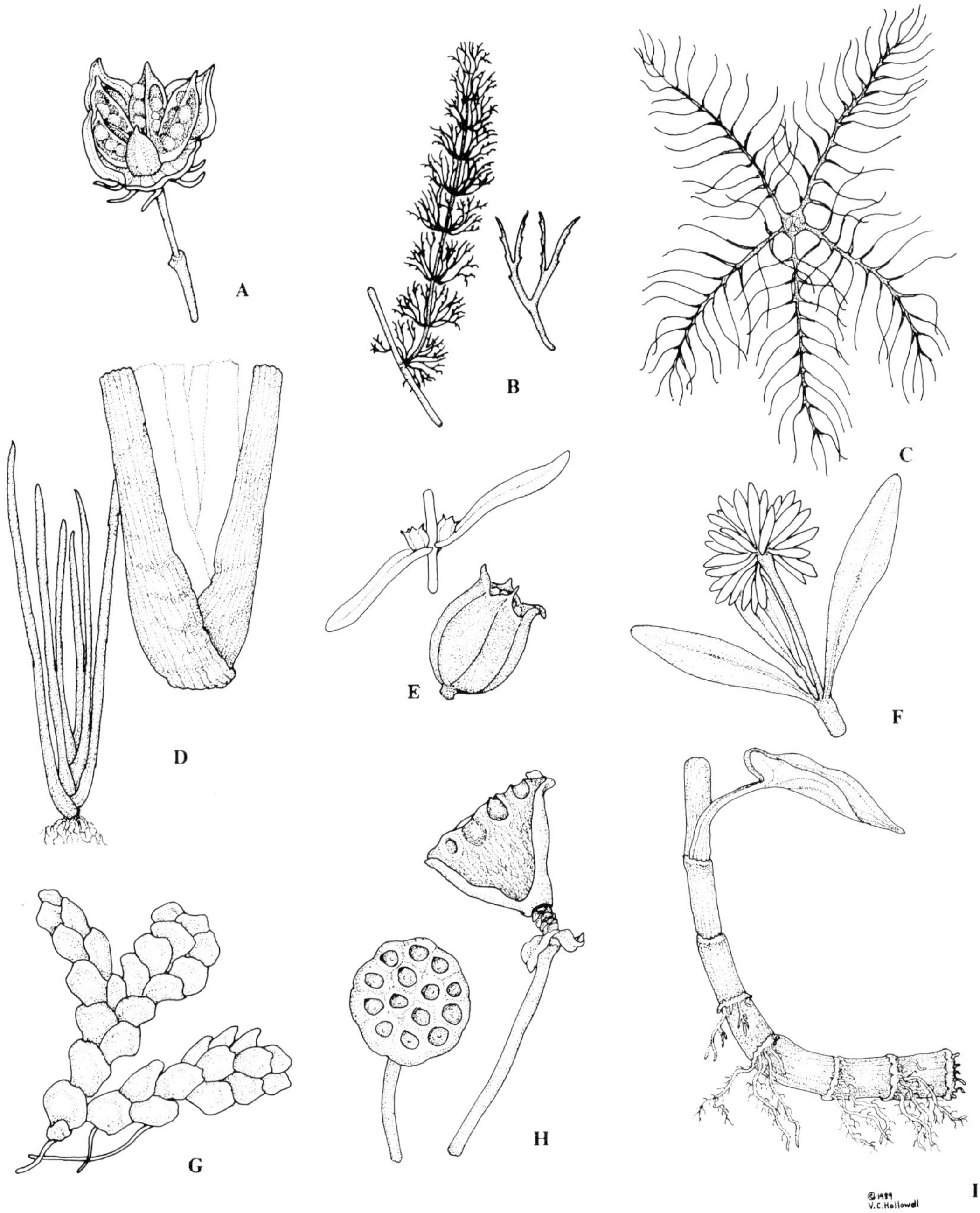

PLATE II.
A. *Hibiscus moscheutos* dehiscent capsule with persistent sepals and bracts.
B. *Ceratophyllum demersum* stem bearing whorls of leaves with dichotomously branched segments.
C. *Myriophyllum heterophyllum* whorl of leaves with deeply dissected and filiform segments.
D. Distichous *Xyris iridifolia* linear leaves with parallel venation.
E. *Rotala ramosior* globose capsules in axils of opposite sessile leaves.
F. *Alternanthera philoxeroides* head of flowers and opposite entire leaves with midribs.
G. *Azolla caroliniana* orbicular leaves in imbricate arrangement.
H. *Nelumbo lutea* indehiscent nut-like fruits embedded in receptacle.
I. *Saururus cernuus* rhizome with internodes and nodes, roots produced only at nodes.

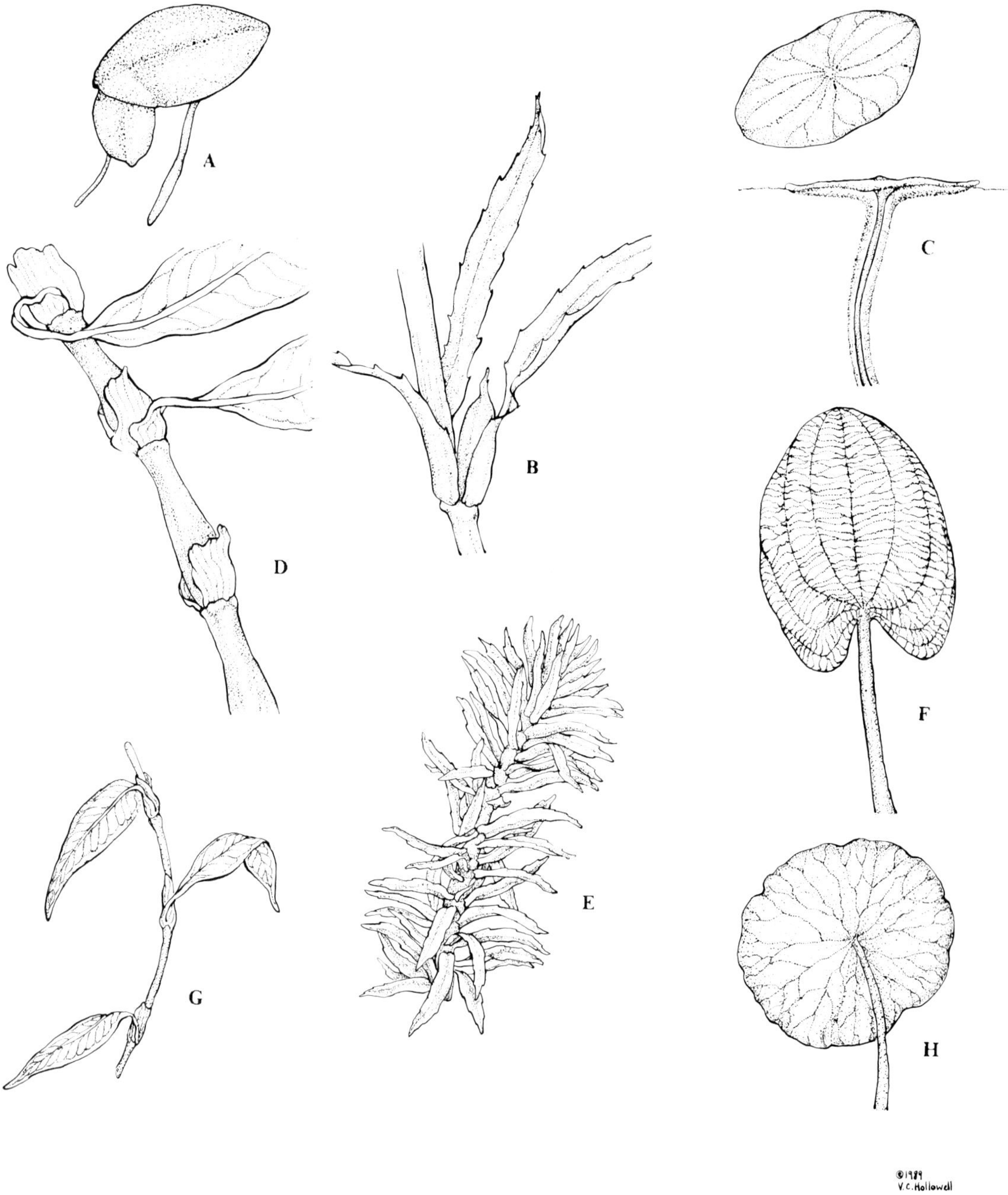

PLATE III.

A. *Lemna* floating leaves with lenticular shape, each bearing a solitary root.
B. *Najas* linear leaves with finely spinulose toothed margins and sheathing bases, enclosing an axillary and indehiscent one-seeded fruit.
C. *Brasenia schreberi* leaf with mucilage covering submersed portion of blade and petiole.
D. *Polygonum lapathifolium* nodose stem with sheathing ocrea at nodes.
E. *Egeria densa* stem with whorls of oblong leaves with acute tips.
F. *Echinodorus cordifolius* ovate leaf with obtuse tip and cordate base.
G. *Polygonum lapathifolium* alternate leaves with pinnate venation.
H. *Hydrocotyle umbellata* orbicular, peltate leaf with crenate margin.

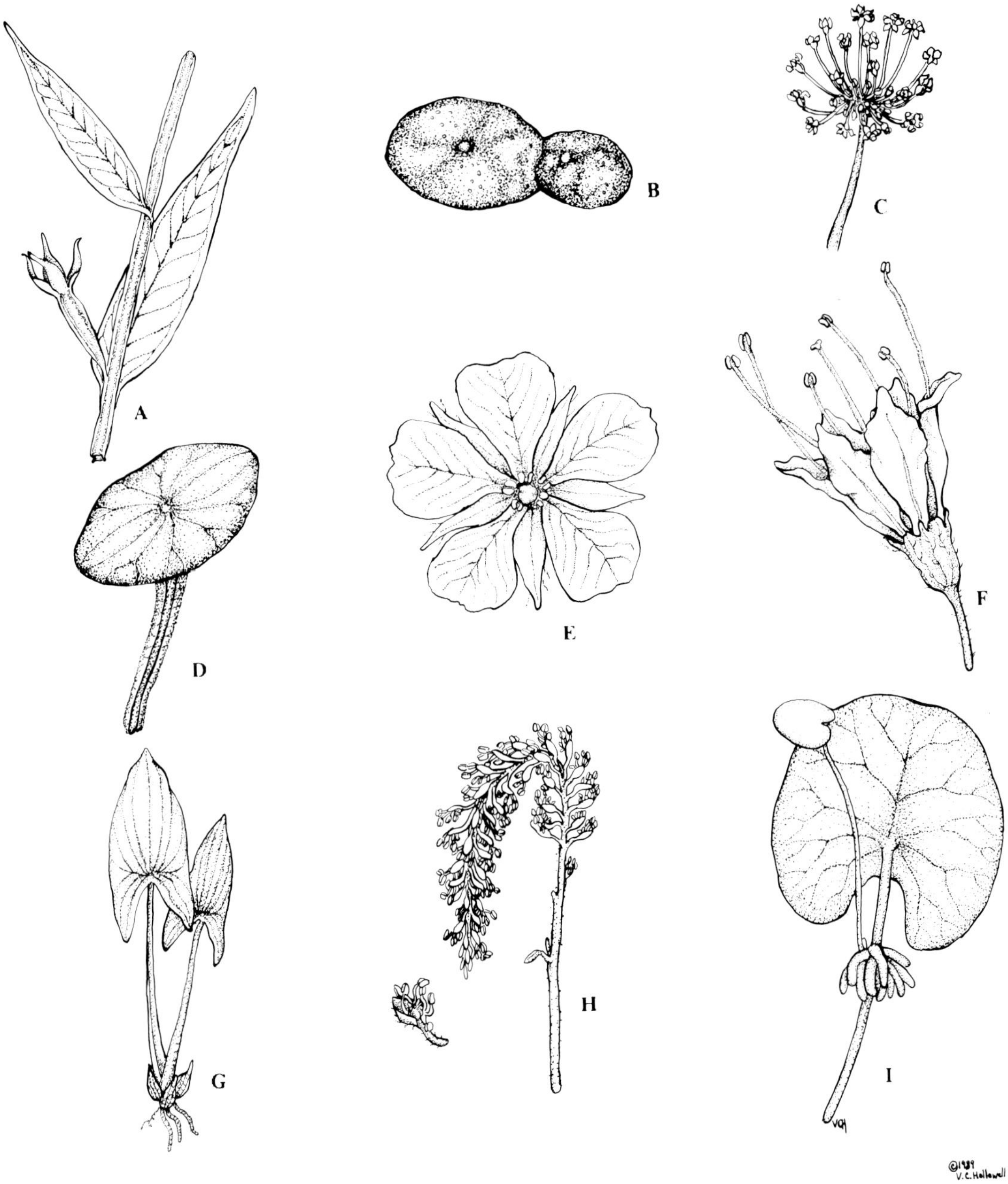

PLATE IV.

A. *Ludwigia decurrens* ovary below four sepals and in axil of lanceolate leaf.
B. *Wolffia* punctate plants with central conical papillae.
C. *Hydrocotyle umbellata* umbel of pedicelled flowers on a solitary peduncle.
D. *Brasenia schreberi* peltate leaf with entire margin.
E. *Ludwigia leptocarpa* perfect flower with perianth, consisting of separate sepals and sepa rate clawed petals.
F. *Decodon verticillatus* perfect flower with fused calyx, separate obovate petals and exserted stamens.
G. *Sagittaria latifolia* leaves with long petioles and blades with basal lobes and palmate venation.
H. *Saururus cernuus* raceme of pedicelled perfect flowers lacking perianths.
I. *Nymphoides aquatica* reniform leaf with tuberous roots at petiole base.

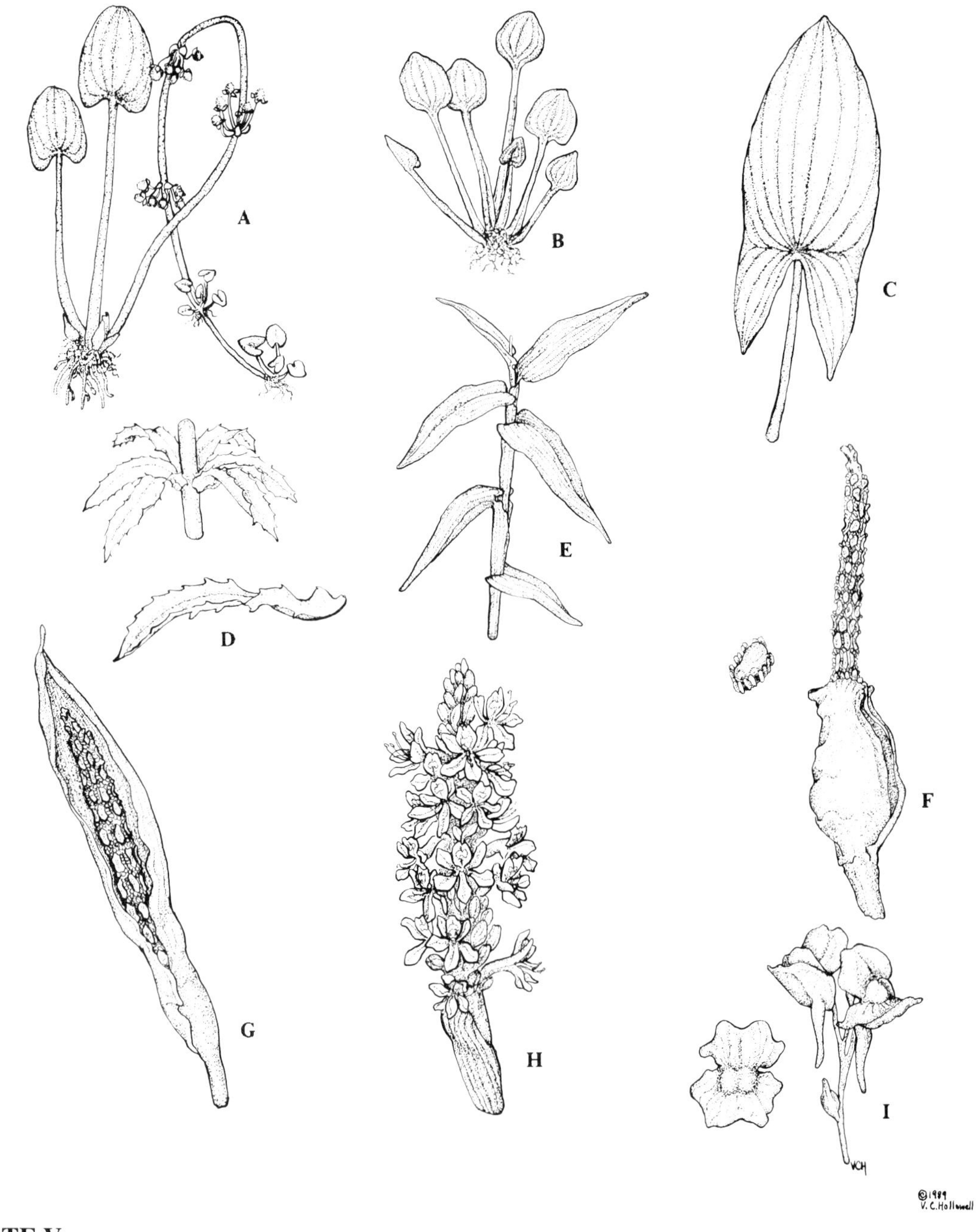

PLATE V.

A. *Echinodorus cordifolius* repent scapes flowering and forming young plants at the nodes.
B. *Limnobium spongia* rosette of long-petioled leaves.
C. *Sagittaria latifolia* simple leaf with pointed sagittate lobes and basal sinus.
D. *Hydrilla verticillata* whorl of sessile spinulose leaves with minute spine-tipped teeth along lower blade midribs.
E. *Murdannia keisak* leaves clasping and sheathing stem.
F. *Peltandra virginica* monoecious spadix with spathe partially removed to reveal upper staminate and imperfect flowers.
G. *Peltandra virginica* elongate spathe enclosing spadix.
H. *Pontederia cordata* spike of sessile perfect flowers.
I. *Utricularia* flowers with two-lipped corolla and spurs on lower petals.

PLATE VI.

A. *Decodon verticillatus* stamens of slender filaments topped by anthers dehiscing pollen from longitudinal slits.
B. *Egeria densa* staminate flower exserted from an axillary spathe.
C. *Potamogeton nodosus* stem with elliptic leaf blades and long stipules at nodes and petiole bases.
D. *Limnobium spongia* stolon producing young fibrous-rooted plants at nodes.
E. *Hydrilla verticillata* turion with recurved bracts.
F. *Hydrilla verticillata* slender rhizome producing late-season tuber.
G. *Ceratophyllum demersum* whorl of deeply dissected leaves with irregularly serrate outer leaf margins.

Index